The
RESCUE
EFFECT

The
RESCUE
EFFECT

THE KEY TO SAVING LIFE ON EARTH

Michael Mehta Webster

Timber Press
Portland, Oregon

To Avani, for believing in me.

Published in 2022 by Timber Press, Inc.
Hachette Book Group
1290 Avenue of the Americas
New York, NY 10104
timberpress.com

Timber Press, Inc., is a subsidiary of Workman Publishing Co., Inc.,
a subsidiary of Hachette Book Group, Inc.

Printed in the United States of America
on paper from responsible sources
Text and jacket design by Faceout Studio
Text is set in Arno, a typeface designed
by Robert Slimbach in 2007.

ISBN 978-1-64326-149-2
Catalog records for this book are available from the
Library of Congress and the British Library.

Contents

Introduction

ON 20 FEBRUARY 2016, Tropical Cyclone Winston became the most intense cyclone ever measured in the Southern Hemisphere as it passed near the island of Namenalala, Fiji, devastating the island's surrounding coral reefs. Winston's winds, with gusts estimated to be as high as 225 miles per hour (360 kmph), generated waves four stories high, which shattered fragile coral fingers and steamrolled the reef with dislodged coral boulders. Wind, waves, and rain stirred up sediments underwater and eroded the nearby islands, which then flooded the reef with silt and pollution.

Exactly four years later, on an unusually calm day, I arrived at Namenalala by boat. I had been working to save coral reefs for years and knew of countless examples of reefs that were struggling to recover after natural calamities such as cyclones. Undisturbed reefs are typically covered with shrub-sized corals that aggressively compete for access to the sunlight they need in order to thrive. In the same way that trees build the forest, these corals build the reef, providing homes to untold numbers of sea creatures. When the corals disappear, the whole ecosystem shifts. A healthy reef that was once bustling with movement, color, and noise in every direction—like a marine version of Manhattan's Times Square—becomes empty, dull, and quiet, like the industrial district of a declining city.

As I snorkeled to the edge of the island's fringing reef, I saw a thicket of small, young corals that ranged from the size of lemons to cantaloupes. It reminded me of a forest that is recovering after being logged, with densely packed young trees scrambling to outgrow one another to hold their places in the light. This reef was unmistakably returning—and swiftly for a coral reef—to its prestorm state.

Why were these corals coming back so quickly? Namenalala's reefs were being seeded by baby corals that drifted in from nearby reefs that were spared the cyclone's devastation. This process of one reef being rescued by others is an example of what I call *the rescue effect*. More broadly, the rescue effect is nature's innate ability to help groups of organisms persist during hard times. Like a thermostat starting the air conditioning when a room gets too warm, the rescue effect automatically turns on when a population is stressed or declining.

In total, at least six different processes contribute to the rescue effect.

- **DEMOGRAPHIC RESCUE** occurs when new individuals immigrate to a small population of organisms—like the baby corals that drifted in to rescue Namenalala's reefs—to provide a numerical boost of organisms that prevents them from going extinct.

- **REPRODUCTIVE RESCUE** occurs when the group of organisms' reproduction and survival rates increase in uncrowded conditions, which increases the population size. Reproductive rescue often takes the form of a baby boom that happens when small populations experience low competition with one another for resources such as food and space.

- **GENETIC RESCUE** occurs when immigrants bring new genetic diversity to a small population, helping it overcome genetic disorders.

- **PHENOTYPIC RESCUE** occurs when an organism adjusts its physiology, outward appearance, or behavior—collectively, its phenotype—to cope successfully with changing environmental conditions.

- **GEOGRAPHIC RESCUE** occurs when a species successfully relocates to a new geographic location after environmental changes render its old location unsuitable.

- **EVOLUTIONARY RESCUE** occurs when organisms evolve, through survival of the fittest, to be able to persist under newly stressful conditions.

The rescue effect is the product of all these processes, which typically interact with one another to help a species persist. There is no purpose or plan behind the rescue effect; in other words, nature is not trying to rescue a species. Instead, the rescue effect is the result of a combination of useful traits that have been accumulated by organisms over eons, plus serendipitous chance events.

Sometimes, the rescue effect is not enough to help species adjust to a changing world. When this happens, humans have an increasingly powerful toolkit for purposefully boosting each of the six underlying processes, thereby helping species persist. Some researchers are even developing an entirely new process to try to bring extinct species back from the dead, Jurassic Park style, which I call *resurrective rescue*. Resurrective rescue occurs when humans use the tools of selective breeding and genetic engineering to re-create approximations of species that have gone extinct.

In the chapters that follow, the processes that create the rescue effect are illustrated as parts of stories about how species—such as tigers in the jungles of India, cichlid fish in Africa's Lake Victoria, and mountain pygmy-possums in the snowy mountaintops of southeastern Australia— are keeping up with a changing world, either on their own or with the help of people.

PERSISTENCE OR EXTINCTION

All living things on Earth, whether they are trees, birds, salamanders, or mushrooms, can trace their origins along an unbroken chain of ancestors going back to the rise of life on Earth at least 3.5 billion years ago. It's difficult for people to understand the concept of billions, or even millions, of years. After all, our recorded history spans only about 5000 years, about one seven-millionth of the time passed since life arose. But every species on Earth today represents a lineage that managed to win the struggle for life, over and over and over, for this almost unimaginably long time.

Many organisms that once thrived on Earth are now extinct, meaning that no individuals of the species exist anywhere on Earth. For example, there are no dinosaurs outside my window, grazing on the bougainvillea. Indeed, most kinds of life that exist in the fossil record are gone because the last of their kind died out or they evolved into new forms. The birds, which are visiting the feeder in my yard as I write, are examples of organisms that survived by evolving from their now-extinct dinosaur ancestors.

Extinction is a natural occurrence that has been happening since life began. In fact, scientists estimate that more than 99 percent of the species that have ever lived are now extinct. The list of extinct organisms includes some that are familiar—such as saber-toothed cats, dodo birds, and woolly mammoths—but also millions of others. And extinction rates are increasing, with most recent extinctions—such as passenger pigeons and Tasmanian tigers—being caused directly or indirectly by people. Extinctions are now happening so frequently that we are entering what scientists call a *global mass extinction event*, when many species disappear over a relatively short time.

Mass extinctions have occurred multiple times in Earth's history. Perhaps the most notorious example marked the end of the Mesozoic Era, when the dinosaurs disappeared. Scientists believe that this mass extinction was caused by a giant meteor that crashed into Earth near the northern edge of the Yucatan Peninsula in Mexico. The crash itself probably shook the Earth with the equivalent force of a magnitude 11 earthquake, enough to trigger aftershocks and volcanos worldwide. One of the biggest threats to life was the dust and steam that the collision injected into the atmosphere; dust probably fell to the Earth for months to years, and sunlight was blocked or diminished for longer, creating something akin to a nuclear winter. Without enough sunlight, plants died and food webs collapsed worldwide. Many of the dinosaurs and other animals that were not killed by the immediate impacts of the meteor probably starved to death.

Today, extinctions are not caused by one large event like a meteor hitting the Earth; they more often result from a thousand human-inflicted cuts. We are harvesting species for food, sport, or materials, sometimes at rates that are unsustainable. We are transforming the Earth to suit our short-term needs when we cut down forests to grow crops or to make pastures for cattle, thereby altering or destroying the homes of forest species. We are introducing species to new places where they compete with, eat, or spread diseases to local species. Finally, by filling the atmosphere with greenhouse gases, we are creating a global chain reaction that is altering the climate—including temperature, rainfall, fire risk, and ocean chemistry—for every species simultaneously. With the world changing so quickly, some species are understandably struggling to keep up.

By one estimate, extinction rates today are 1000 times higher than before humans took over the world. This means that for every single species that would have naturally gone extinct on its own, roughly 999 more are now disappearing because of us. Although the growing rate of extinction is cause for alarm, the actual number of extinctions to date remains low relative to the diversity of life on Earth. For example, one group of scientists have estimated that the background extinction rate—the number of species that would typically go extinct if people were not changing the planet so quickly— is about 0.1 extinction per million species per year. With roughly two million known species on Earth, that means that prior to the rise of human populations, about one species would go extinct every five years. In contrast, we are currently losing something like 200 species every year as a result of human activity. Though many conservation advocates would argue that even a single human-caused extinction is too many, from a purely numerical perspective, 200 is only a 0.01 percent loss of global species diversity per year. At this rate, it would take a century to lose 1 percent of the known species on the planet.

It's worth pausing a moment to ask, with everything humans have already done to transform the world in the past few millennia, why is the extinction rate still so low? The answer is that, for most species, the rescue effect is working, so far. As their environment has changed, most species have relied on natural systems to kick in and help them adjust. This is great news for conservation, because it means that a large majority of species are adapting to a changing world on their own and, at least for the time being, won't need any special help from people. Accordingly, people can focus on helping the relatively few species that we know are struggling the most to keep up. Fortunately, in most of these cases, we still have time to intervene before they are lost.

This book dives deeply into the stories of species and ecosystems that are adjusting to our changing world. In some cases, the rescue effect is proving strong enough, even when the odds initially seem long. In other cases, the rescue effect has been overwhelmed, and without concerted efforts by people to give nature a boost, extinction is the most likely outcome.

As observers to—and the underlying cause of—the decline of species that are struggling, we get to choose whether and how to help them persist. Today, people are already making these kinds of decisions in ways that are pushing the limits of science and technology, while raising new questions about conservation policy and ethics. For example, what species should we prioritize to rescue? Is it acceptable to let other species go extinct? And if we do intervene, should we use tools and emerging technologies that create their own risks? The answers to these questions won't always come easily. But there are good reasons to be optimistic, because everywhere we look, we can see evidence that nature can rescue many species from extinction, and when nature alone is not up to the task, we can help.

CHAPTER 1

Son of Panna

"MADAM, THE TIGER WILL NOT ATTACK," our guide said for the second time.

"But what if it does? What's our plan?" returned my wife, Avani, for the third time.

We were on a safari drive in Bandhavgarh National Park, in the central Indian state of Madhya Pradesh. I had dreamed of going on a safari all my life, so when we made plans to visit Avani's parents in Delhi in 2017, I requested that we make a side trip to one of India's many nature reserves. If we were lucky, we might encounter a tiger.

We asked family and friends for recommendations about where we would have the best chance of seeing tigers. After all, we had heard many stories of people who had gone on safari without spotting a single one of these elusive cats, and we wanted to choose our destination to maximize our chances of success. The consensus view was that we should go to Bandhavgarh, which had a promising combination of high tiger densities and many open clearings and water holes where they could be regularly seen.

The morning of our trip, we got up early to take a domestic flight from Delhi to Jabalpur. From Jabalpur, we rode five hours in a car through small towns and villages separated by agricultural fields. It was late June, and the summer monsoons were just starting, so many of the fields were being prepared for planting with wooden ploughs pulled by teams of oxen. This land had once been tiger habitat, but as human populations have swelled in India over the centuries, more and more land has been cleared for the cultivation of crops, the building of cities, and the pursuit of other human priorities.

When we arrived at our hotel outside of Bandhavgarh National Park, it had already been a long travel day, so I think the hotel staff was surprised when we asked if we could go out for our first safari drive that evening. Bandhavgarh is intensively managed to protect its wildlife, including its big cats. With so many people wanting to visit, to keep the park from becoming overly congested, officials have a limited entry system that requires a unique permit every time someone enters the park. The permits for the most desirable core areas of the park were already taken for the day. Fortunately, permits were still available for the buffer zone, a less-visited ring around the park shared by local villagers, their livestock, and wildlife that strays out of the park.

When we arrived in the buffer zone, we learned about a recent leopard kill just up the road and headed there to see if the leopard might come back to feed. As we drove along the red dirt roads, we stopped periodically to see wild boar and spotted deer, both potential food for tigers. Before arriving at the leopard kill, however, I saw something on a side road that made me freeze. At first, I was so shocked that I had trouble spitting out the word, but I finally managed to say, "Tiger!"

I had imagined that, if I were lucky enough to see a wild tiger, it would be creeping through dense jungle, mostly obscured from sight. This tiger was strolling in the middle of the road, as calmly as the ever-present cattle in the streets of Delhi. He didn't pay any attention to us, but instead stopped to spray a tree with urine, marking his territory.

I had seen plenty of tigers in zoos, so I had a good idea of their size, but I was awed by this tiger's physique. He was a living feline version of Michelangelo's *David*, with perfectly formed muscles in his shoulders, neck, and limbs that protruded through his velvety orange and black fur as he took each firm step. It dawned on me then that I had seen only the "dad-bod" version tigers that spend their days in relative leisure in zoos, waiting for the next meal to be delivered. Dad-bod tigers are formidable enough, but this animal was akin to a professional athlete, capable of downing an animal five times his own size with just the strength of his limbs, paws, and jaws, and then dragging it away in his teeth.

To our surprise, this tiger, which we later learned had been named Panna Lal by the guides, started ambling directly toward our jeep. Our guide didn't seem the least bit worried about this change of events and casually began backing up the jeep to put some distance between the tiger and us. I was so absorbed in taking photos that I wasn't thinking about any risks. But Avani was. Our safari jeep was open to the elements, with no windows other than the windshield, and no metal frame or roof. In other words, there was nothing between us and Panna Lal, should he decide that we looked tasty. Furthermore, our guide carried no weapon.

Once we had backed up a few hundred feet, our guide put the jeep in park. Panna Lal was still walking straight toward us. Avani, who is a law professor, decided to continue pressing her case. "Seriously, if the tiger attacks, what do we do?" By this point, she had worn our guide down. He clearly didn't want to answer this question but realized that it was time to level with her.

"Madam, if the tiger attacks, there is nothing we can do. The tiger will do what it wants."

A human, unarmed, is woefully inadequate to resist a tiger. The largest cat in the world, an adult male tiger can weigh more than 500 pounds (227 kg). In addition to their extremely powerful muscles, these tigers have claws and canine teeth that each measure up to 3 inches (8 cm) long—that's probably about the length of your longest finger. Tigers are not distance runners, but instead are ambushing hunters that explosively pounce on their prey. In a single leap, they can travel 30 feet (9 m), and as the cat once again drew nearer, he was easily close enough to land cleanly in our jeep.

A few years ago, I'd seen a video of a tigress with cubs. Rangers riding on the backs of elephants were attempting to tranquilize and relocate her and her family to reduce conflicts with local villagers. The rangers instead triggered her instinct to protect her cubs. She attacked, leaping from the tall grass, mouth wide open, left paw held up to swipe at an elephant driver on the back of an elephant. Asian elephants stand 8 to 10 feet (2.5 to 3 m) tall, and yet the tigress had no trouble delivering a hard smack to the driver atop the elephant. It's hard to watch the short video and not be in awe of her power, speed, and protective instincts.

PANNA LAL

At the time we encountered Panna Lal, I didn't know anything about his backstory or the unexpected ways he would come to epitomize the rescue effect. The first thing I learned about this remarkable cat was from our guide, who explained that Panna Lal was new to Bandhavgarh and had arrived only months earlier. Initially, no one knew where he came from. But an informal team of photographers and wildlife managers had solved the mystery just a few weeks before our arrival.

Tigers have distinctive striping patterns that can be used like fingerprints for identification. By comparing photos of this new tiger with pictures of tigers taken elsewhere, park authorities identified him as a three-year-old cat that had disappeared from Panna Tiger Reserve, about 90 miles (140 km) to the northwest, in late 2016. Months later, he arrived in the buffer zone of Bandhavgarh. He was thereafter given the fitting name of Panna Lal (Hindi for son of Panna).

In tiger society, young tigers leave their mother's territory at about age two. Their mother has cared for them constantly, from the time they were defenseless cubs until she has taught them everything she knows about hunting and survival. At that point, it's time for the adolescents to leave and fend for themselves.

The territories of tigresses are usually nested within the larger territory of a dominant male. The male's territory may span that of three or four females, with whom he will mate, producing as many cubs as possible. He also defends his territory from rival males, who may attempt to unseat him and take over. If successful, the challenger may start his reign by killing all his predecessor's cubs so that the females will go into heat and he can start fathering cubs of his own. This is the tough world into which Panna Lal was born in May 2014.

According to R. Sreenivasa Murthy, who was the field director of Panna Tiger Reserve when Panna Lal was born, this tiger had nearly died when he was a cub. Apparently, when he was a little more than a year old, he got into a fight with one of his siblings and was left with a large gash under his chin. When Murthy heard the report that this tiger was wounded and could die without medical attention, he thought, "I had to secure my kitty at any cost." So, along with a rescue team, he jumped into action. The team tranquilized Panna Lal and sutured his wound, and he was able to recover fully.

By the time Panna Lal was two years old, he would have been starting to wander farther from his mother to hunt and kill his own food. He was getting ready to set off on his own. This would have suited his mother just fine, because she needed her full territory to feed herself, and it's likely that she had another litter of cubs on the way.

Like most tigers nearing adulthood, Panna Lal set out to find a territory of his own, but he didn't find what he was looking for within the park where he was born, so he searched more widely. According to Murthy, Panna Lal was spotted on a camera trap in the wooded, southeastern edge of the park as he started his journey to Bandhavgarh. These forest-covered hills probably served him reasonably well, giving him cover and providing potential prey. After that, Panna Lal decided to cross densely cultivated areas, where he was more likely to encounter people.

During Panna Lal's journey, Murthy recalls receiving several calls from villagers, with messages like, "Your tiger is here in my family farm." Murthy advised them to keep their distance to avoid any conflict. Tigers are legally protected in India, but it would be hard to fault anyone for wanting to protect their families or livestock from such a dangerous predator. Imagine seeing a tiger wandering through your yard. I have coyotes in my neighborhood in California, and I worry about whether they could harm our youngest children. Tigers are easily ten times the size of an adult coyote.

As Panna Lal explored the countryside, he would have had to eat regularly. In cultivated lands, his easiest options would have been livestock, especially cattle and domesticated water buffalo. For the livestock owners, this would have meant the terrifying prospect of having their animals—from which they make their livelihood and feed their families—attacked, killed, and consumed.

At some point, Panna Lal developed a taste for domesticated animals. According to Murthy, Panna Lal's mother had similar preferences, so he may have developed his culinary palate while still a cub. By the time I saw him at age three, our safari guide told us that the tiger already had a reputation for disregarding the abundant deer and pigs in his territory. Instead, when he was hungry, he would attack a relatively pliant domesticated cow or buffalo grazing in the buffer zone or in a nearby village. With this lifestyle, he found the buffer zone an ideal location to establish his territory. After all, he was still growing, and he needed to bulk up before he could start challenging adult males for access to mates.

GLOBAL CRASH

Panna Lal was born into a much more constrained and dangerous world than his ancestors. Just 200 years ago, tigers roamed across a giant swath of Asia, throughout the Indian subcontinent, south into Malaysia and Indonesia, and sweeping northeast through China and Russia. A separate population inhabited Central Asia, ranging from Georgia in the west to Mongolia in the east. Scientists have used DNA samples to identify nine distinct tiger populations, from the large Amur tigers in northeastern China and into Russia, to the relatively diminutive but now extinct Javan and Balinese tigers that lived in the jungles of Indonesia. Bengal tigers like Panna Lal are the type that live in India. Given enough time, these separate populations may have branched into new species, increasing the diversity of the world's big cats. But instead of continuing to diversify, tigers are rapidly disappearing.

By one account, tigers have lost 93 percent of their historic territory primarily as a result of human land use. That means that the remaining wild tigers in the world are confined to just 7 percent of the places they roamed in the recent past. (For comparison, the states of Colorado and Nevada combined are about 7 percent of land area of the forty-eight contiguous United States.) Moreover, the remaining tiger habitat is not concentrated in one area; instead, it's highly fragmented, like an archipelago of small islands isolated in a sea of villages, croplands, and cities.

In addition to losing most of their historical lands, tigers are killed by people. In some cases, tigers are killed for self-protection. After all, who wants to share their land with a tiger? Such conflict with people would have been a very real risk for Panna Lal on his journey from Panna to Bandhavgarh. It's also a risk for any tiger that lives at the edge or wanders out of parks.

But many tigers are killed for a more insidious reason: there's an international market for tiger parts. Humans have long sought to possess parts of organisms that they admire and fear, such as elephant tusks, shark teeth, and eagle feathers. This desire is often related to a deep-seated belief that the owner will supernaturally take on some of the possessed organism's traits. To feed this desire, a multi–billion dollar worldwide market for wildlife parts is fueling overharvest and thus threatening the extinction of many species, including tigers, rhinos, gorillas, and elephants.

There is a market for nearly every part of a tiger. People have long used the furs of big cats to project wealth and power—whether it's a high-society socialite in times past with a leopard fur coat or a big game hunter with a stuffed lion. While wearing big cat fur and trophy hunting may seem abhorrent in contemporary Western culture, the fur is still often the most valuable part of a poached tiger. Tiger teeth and claws are used to make amulets and jewelry. Their meat is served as a luxury food. Their organs and bones are used to produce an extraordinary array of unproven folk medicines, such as tiger-bone wine to reduce inflammation, tiger eyes to help cure epilepsy and cataracts, tiger penises to treat impotence, and tiger stomachs to help relieve indigestion. A dead tiger, once dismembered and partitioned into is various products, can be worth a lot of money. It's no wonder, then, that when given the opportunity, some people will kill tigers for profit.

Many laws and regulations have been enacted with the aim of stopping tiger hunting in the countries where they are still found. The international trade of tiger products is also banned through the Convention on International Trade in Endangered Species (CITES), which went into effect in 1975 and was ratified by India a year later. However, today there remains a thriving black market for tiger parts, largely for customers in China and parts of Southeast Asia, which fuels ongoing tiger poaching operations. Because wild tigers live in remote locations, poachers can often hunt and snare them undetected.

As a result of habitat loss and poaching, the global wild tiger population has crashed, from around 100,000 animals in 1900, to about 4500 in 2008—a decline of about 95 percent during that time. Most of the holdouts are Bengal tigers in India, like Panna Lal, which number around 3000; the rest persist in small pockets of tiger habitat across Indonesia, Southeast Asia, and Russia.

PANNA TIGER RESERVE

The current state of tigers in India is cause for both urgency and optimism. India's tigers have experienced the same pressures that affect their kin elsewhere. However, there are signs that the population has passed through its nadir and is starting to recover. Perhaps no other place in India shows this turnaround more clearly than Panna Lal's birthplace, Panna Tiger Reserve.

In the 1950s and 1960s, India's Bengal tiger population was declining. By the 1970s, it was clear that without action, India's national animal would go extinct in the wild. The government finally decided to act by passing national legislation in 1972 to launch Project Tiger, which promoted the designation of tiger reserves, where tigers and other wildlife were given free reign and protection from habitat destruction and hunting. According to Bittu Sahgal, editor of the magazine *Sanctuary Asia*, Project Tiger has been "one of the most successful rewilding exercise in the past 100 years." The term *rewilding* refers to returning lands that have become tame, often because of the loss of large animals, to something resembling their wilder past. To date, Project Tiger has succeeded in reversing the decline of tigers in India and securing many of the parks where tigers currently roam. But this overall success has come with some significant bumps in the road.

Panna National Park was created in 1981 and designated as India's twenty-second tiger reserve under Project Tiger in 1994. By 2001, Panna National Park looked like a success story, with a respectable population of about thirty-one tigers. However, the park's initial success would prove short lived, as Panna's tigers began to disappear.

In 1996, wildlife biologist Raghu Chundawat established a long-term tiger research program in Panna that would end up documenting the decline of the park's tigers. Interested in the details of tiger behavior and ecology, he conducted studies of individual tigers. His team attached radio collars to the animals to track their movement, and over the years, he followed these tigers throughout their lives, witnessing births and deaths, family dynamics, and territorial disputes. Team members rode on the backs of elephants, which many tigers will ignore, to watch and document tigers going about their daily business in the jungle.

Chundawat noticed that something started to change in Panna's tiger population in 2002, when two breeding-age females disappeared: one tiger's body was found in a poacher's snare, and the other was never recovered. In 2003, a large male was found dead in a well. Over the next two years, more tigers disappeared in the park. Chundawat was alarmed, writing, "We had known some of them from the time they were born. Losing them was like losing a part of one's family."

Forty-two villages are located in and around Panna Tiger Reserve, and most villagers have found ways to coexist relatively peacefully with tigers. Villagers watch and guard their livestock and rarely encounter tigers. When people in these villages do have conflicts with tigers, they may fight back in various ways. One way they attempt to rid themselves of a problematic cat is to set out the poisoned carcasses of dead livestock to entice a tiger to eat the tainted meat. Villagers may also accidentally kill a tiger while trying to illegally snare another animal, such as a deer, for food.

Because villagers mainly interact with tigers on the periphery of the park, the tigers in the interior core of the park should have been relatively safe from these kinds of risks. However, in the early 2000s, members of the Pardhi tribe were venturing deep into the park to hunt Panna's tigers for profit. Pardhis have long been known as preeminent hunters and trackers in India; their tribal name is derived from the Sanskrit word *paparddhi* (hunting). They once enjoyed relatively high social status and served as hunting guides for various rulers in India, including the Mughals and the British. That all changed when India's British rulers began systematically disempowering many groups, including the Pardhi, by branding them as habitual criminals in the Criminal Tribes Act of 1871. The Pardhi and their descendants were permanently presumed to be engaging in criminal activity. While the British ultimately left India and the Criminal Tribes Act was eventually rescinded, its stigma has never entirely been erased in India. Many Pardhis remain socially and economically ostracized, making it difficult for them to develop new economic opportunities.

At the time Panna's tigers were declining, a small group of Pardhi families were living nomadically in and around the park. I spoke with Bhavna Menon, a program manager with the Last Wilderness Foundation, who works with this Pardhi community. She said that these families had no permanent home, but instead would move temporary camps from place to place in search of new areas to hunt. At that time, some Pardhi hunters were generating income by supplying wildlife traffickers with a variety of animals poached from the park, including tigers and leopards. These traffickers were middlemen who would pay the Pardhi for their kills and then move the wildlife products through an organized crime network, first within India, and then internationally.

Increased poaching in Panna coincided with the rise of a new market for tiger furs in Tibet. Tibetans wore the expensive skins for fashion and to signal wealth. In 2005, Belinda Wright of the Wildlife Protection Society of India decided to visit Tibet to see for herself what was happening there, later writing, "What we found, on our very first day there, was beyond my wildest nightmares. Hundreds of people were openly parading and dancing in fresh tiger and leopard skins. In Lhasa we found tiger skins for sale, openly and on a large scale. All the traders we spoke to said that the skins had come from India."

Meanwhile, back in Panna, wildlife biologist Chundawat was so alarmed by the uptick in tiger deaths that he notified the park's managers, expecting that they would spring into action. Instead, they insisted that everything was fine in the park. According to Chundawat, park managers then tried to silence him by temporarily restricting his ability to enter the park to document the declining tiger populations.

By 2005, Chundawat had regained access to the park and concluded that, since 2002, "Panna had lost nine breeding tigers out of eleven known to us. This represented almost the entire breeding population of the Reserve." At the same time, he argued that a full-blown cover-up was underway and extended all the way to the top, with Project Tiger Director Rajesh Gopal writing that park staff had "estimated more than 30 tigers" still inhabited the reserve. Chundawat says that this estimate was fabricated by counting the same tigers repeatedly. In one case, park staff counted "13 different adult tigresses within the territory of one tigress." Such a density of tigers would constitute a biological miracle that goes against everything known about tiger territoriality. If a female wouldn't suffer a single rival in her territory, she certainly wouldn't allow twelve! The simpler explanation was that the tigers of Panna were in steep decline as a result of poaching and the managers were cooking the books to cover their tracks.

The cover-up finally crumbled in January 2009, but by then it was too late. A new comprehensive tiger survey had just been conducted, complete with extensive camera trapping. Researchers couldn't find a single tiger in Panna Tiger Reserve. In less than seven years, poachers had managed to kill every last animal.

The loss of Panna's tigers was an enormous embarrassment to the government of India. It didn't help that it came on the heels of another high-profile tiger extirpation in Sariska Tiger Reserve, where tigers were eradicated by poachers in 2004. The crashing tiger populations in India were looking more and more like the result of inept mismanagement. The countrywide tiger population was reinforcing this image: in 1973, when Project Tiger was launched, India had an estimated 1800 tigers. After more than thirty years of strict protections, India's tiger population continued to decline, reaching an estimated 1411 tigers in 2006.

Although there are reasons to believe that these population estimates were imperfect, it was clear that many tiger reserves were being overwhelmed as poaching for wildlife products ramped up in the 1990s. Panna Tiger Reserve's tiger extirpation was just one more high-profile indication that wildlife managers would have to stop rampant tiger poaching if India wanted to keep its tigers.

The return of Panna Tiger Reserve

Effective wildlife management often involves two very different ingredients: managing wildlife and managing people. In Panna, the biggest problem for tigers was people, and it was clear that the approach of the current management team was failing. As the park was hitting its reputational bottom, the Indian government decided to bring in new park management.

Just months after the last tiger disappeared from the reserve, R. Sreenivasa Murthy was posted as the new field director. He told me that this was a very challenging role to step into. At first, "people were just spitting venom on our faces because we were seen as the primary culprits of losing the tigers in Panna."

Murthy took it upon himself to rehabilitate both park management and the reputation of Panna Tiger Reserve. He told me that his management approach was founded on one basic principle: zero tolerance for any poaching. But before he could make this principle a reality, he had to change the organizational culture of the management staff. He said the enforcement staff had often been pressured to look the other way, or even collude in poaching. He made it clear that, "if you are going to be with the other side, I am not with you." During his tenure, he cleaned house, removing staff members who were not onboard and replacing them with others who would be. He said that four of his uniformed staff were arrested for wildlife offenses and went to jail.

Murthy also got serious about securing access to the reserve. For example, the Ken River on the western side of the reserve could be crossed by boat, giving poachers easy access to the reserve. Murthy added new posts along its shores, where wildlife officers would stand guard to prevent poachers from entering.

He also needed to confront the problem of the Pardhi poaching within the reserve. Conflict between the economic interests of people

and the goals of wildlife management is extremely common. It's sometimes easy to conclude that wildlife parks just need more enforcement
to keep people out, but this simplistic approach can miss the cultural
and historical complexities that cause a group like the Pardhi to poach.
Pardhi children grew up within their tight-knit, nomadic community
with no access to any formal education and therefore no real economic
prospects beyond continuing to poach. Murthy said every time his staff
tried to remove the Pardhi from the park, they'd sneak back in and continue to make a living the only way they knew how, by hunting. In total,
his staff threw them out of the reserve at least thirty-eight times in a
period of six years. Clearly, forcible removal wasn't working and a more
sustainable solution was needed.

Bhavna Menon told me that the forest managers finally tried a different approach: negotiating directly with the Pardhi to try to make a deal.
The forest managers offered to set up schools for the Pardhi children in
nearby Panna City, where the families could also settle down in permanent housing and pursue alternative sources of income. In return, the
Pardhi were asked to stop all tiger poaching. Menon's organization is
currently working with the Pardhi to develop alternative livelihoods,
such as working as nature guides, which enables them to use their skills
at tracking wildlife, as well as engaging in poultry farming and rickshaw
driving. The organization has also been supporting the education of the
Pardhi children with the aim of providing the next generation with a new
set of social and economic opportunities.

Tiger management in Panna even got a boost from the Dalai Lama,
who, after being shown images of Tibetans wearing tiger skins in 2005,
urged Tibetans to "not indulge in such stupid and senseless habits." The
response to the Dalai Lama's words was swift, and when Wright's organization returned to Tibet for further investigation in 2007, they didn't
find a single person wearing tiger skins. Although Tibet was not the only

market for illegal tiger products, turning off the spigot there likely helped curtail demand from places like Panna.

After tigers disappeared from Panna, Murthy encountered intense pressure to rebuild the park's tiger population quickly. As he explained to me, the shame of losing Panna's tigers meant that the organization faced scrutiny regionally, nationally, and internationally. In that kind of environment, he said, "you try to pull up your socks very fast." So, in parallel with getting a handle on poaching, park staff began working on restoring the tiger population. To do so, they looked to demographic rescue.

In biology, demographic rescue occurs when new organisms immigrate into a population, boosting the population size. The tendency of young adult tigers like Panna Lal to explore the landscape for a place to establish their own territory means that some individual tigers are always on the move, looking for open spaces. So if local tiger populations went to zero in a place like Panna Tiger Reserve, it would just be a matter of time before new tigers wandered in. Such immigration can lead to demographic rescue if new individuals arrive from elsewhere to rejuvenate a small or locally extinct population.

Because other tiger reserves in the area were so far away, managers knew that it could take a long time before enough new tigers arrived in Panna on their own. Instead of waiting, they decided to boost immigration by capturing wild tigers elsewhere and relocating them in Panna.

Managers got right to work, capturing one female from Bandhavgarh National Park, another female from Kanha National Park, and a male from Pench Tiger Reserve to relocate to Panna in 2009. Relocation can be very stressful to the tigers: they are shot with a tranquilizer dart, captured, and then released into a foreign landscape to fend for themselves. In many cases, translocated animals die or wander off, sometimes eventually returning to their original homes.

To keep track of Panna's new tigers, managers fitted all three animals with radio collars. The two females remained in Panna, but the male quickly strayed out of the reserve. He was then tracked, tranquilized, and transported back to Panna a second time, where he subsequently stayed.

Reproductive rescue

By the end of 2009, Panna Tiger Reserve had gone from zero tigers to three, and three more animals were reintroduced over the next four years. With poaching diminishing and more tigers being introduced into the park, management had made good on their intention to boost the tiger population and reduce mortality. Since then, Panna tigers have been able to rescue themselves, starting with a baby boom of tiger cubs.

For the new tigers, Panna Tiger Reserve provided an abundance of food, little competition with other tigers for space, and open territories for young tigers when they set out on their own. These are excellent conditions for reproductive rescue, which happens when an unusually large number of young animals survive to adulthood. In a population experiencing such a baby boom, each adult produces more than enough young to replace itself, causing the population to grow. For large predators, tigers have high reproductive rates, with females typically producing two or three cubs every two years. At this rate, a successful tigress can produce a dozen cubs in her lifetime. Under poor conditions, few cubs survive long enough to have cubs of their own. However, in favorable conditions—as was the case when tigers were reintroduced to Panna—a female can rear many cubs that survive long enough to reproduce.

Consider the case of the tigress relocated from Kanha National Park to Panna Tiger Reserve in 2009. She was simply named T2, for Tiger 2, when she arrived in Panna. In October 2010, she had her first litter of three cubs. T2 went on to birth two more litters, totaling at least eight cubs in Panna. At least four of her cubs were surviving on their own at the time of her death around 2014, doubling the population of adult tigers in one generation. When T2 died, some of her offspring were starting to produce litters of their own, which further increased the population. One female from T2's 2010 litter gave birth to four cubs in May 2014, one of which was Panna Lal.

The tiger baby boom in Panna has led to a rapid increase in their numbers. In a recent report by the Indian government, biologists estimated that thirty-one tigers were living in and around the park by 2018, restoring the population to what it was in 2001, prior to the poaching-driven crash. Menon told me that the tiger population in Panna had already reached sixty by 2020, and Murthy thought the 2020 population could be even higher, perhaps more than eighty animals, if the tigers that live on the outskirts of the park are included. The recipe for rescuing tigers in a place like Panna is, at least in principle, pretty simple: start with a few tigers (in this case tigers introduced from other parks), give them plenty of space, and simultaneously curtail external sources of mortality, which in Panna was poaching. Reproductive rescue will take care of the rest.

CONNECTING PARKS

The most tried-and-true tool for wildlife conservation is arguably the nature park, where human activities are minimized, giving nature an unfettered chance to take its course. Although no places on Earth are truly free of human influence today, reducing the local human footprint in a park remains an effective conservation strategy in many situations. As was the case for Panna Tiger Reserve, this strategy, along with an initial intervention by people to reestablish tigers and ongoing reductions in poaching, has led to the big cats' swift recovery.

Nevertheless, the Panna reserve has a fundamental flaw: the park alone is not big enough to host a healthy population of tigers indefinitely. That's because nearby human land development has turned Panna into an isolated island of tiger habitat. Within this island, each tiger requires a lot of space, and they are willing to defend that space with tooth and claw. At a certain point, a landscape can become saturated with tigers defending their territories, making it nearly impossible to squeeze in any more tigers. In ecology, this point of maximum sustained population is known as the *carrying capacity*. At 208 square miles (540 square km), Panna Tiger Reserve may not seem small to a human observer, but it fills up quickly with tigers. For example, female tigers in this kind of dry forest will each defend territories of around 7 square miles (20 square km). This means that, even with perfect management and uniformly good habitat, the park will be able to support only about thirty female tiger territories.

Furthermore, in populations of a few dozen breeding adults, animals can have trouble finding a mate that is not a close relative such as a sibling or a cousin. This results in inbreeding, by which mating with close relatives causes populations to lose genetic diversity and increases their chances for genetic disorders. People instinctively know that inbreeding is a problem, which is one of the reasons why most human cultures find incest so abhorrent. Small, isolated wild animal populations don't have much of a choice: they either inbreed or go extinct. The reserve had reintroduced only a handful of tigers, which created an especially high risk of inbreeding from the start.

The state of Madhya Pradesh, where Panna is located, is home to six tiger reserves with an estimated 403 tigers as of 2018. Some tigers also live outside of the designated tiger reserves, and neighboring states also house tigers, which collectively brings central India's total tiger population much higher. If central India's tigers could act as one large population, rather than a collection of small, isolated ones, they could largely avoid the genetic problems of inbreeding.

Unbeknownst to Panna Lal as he traversed the 90 miles (140 km) between his old home and his new home, he was showcasing how this could happen through genetic rescue. For small and isolated populations, immigrants such as Panna Lal can bring more than just new individuals to the population (demographic rescue); they can also bring new genes. When mating occurs between established residents and newcomers, the whole population's genetic diversity can increase.

Movement between tiger populations would have been common before the landscape was fragmented into isolated parks. But how is a tiger to get from one park to another now? Panna Lal managed to find a way, but he was probably very lucky. Forest managers would like to make it easier for tigers to make journeys like Panna Lal's by establishing wildlife corridors, which are usually thin strips of protected wildlife habitat that connect isolated parks. Corridors are designed to act as wildlife highways, where animals can travel from one park to another without the need to enter areas with concentrated human development. On a map, this often looks like a spider web of corridors connecting scattered, isolated parks. Some corridors already exist in Madhya Pradesh because of existing protected lands, forested river courses, or mountainous areas that have lower human population densities.

In addition, wildlife managers and conservationists are using sophisticated mapping software to identify new places that, if reverted to wild lands, would make it easier for tigers like Panna Lal to travel or live between the reserves. These maps also can help inform whether and where to build new human infrastructure such as highways or railroad tracks.

In addition to bringing in new genes to keep isolated populations genetically healthy, a connected network of tiger populations would make it unnecessary for managers to relocate tigers from one park to another, because the tigers would do it themselves. As Panna gets more crowded, a surplus of young tigers may have an incentive to follow Panna Lal's lead and look for new territories elsewhere. Noradehi Wild-life Sanctuary, about 100 miles (160 km) southwest of Panna, has only a small population of tigers and is connected to Panna by existing wildlife corridors. If a few tigers find their way to Noradehi from Panna, they could bolster that sanctuary's population.

Madhya Pradesh is not the only part of India with a connected tiger reserve network that is large enough to accommodate a healthy breeding population of wild tigers. There are similar areas, for example, in Northern India, including parks across the border in Nepal and in the forested mountains inland from India's southeastern coast. Whether or not these areas will continue to support healthy populations of tigers in the future will be partly determined by the status of their corridors. In Madhya Pradesh, a less tiger-friendly future could include a steady erosion of the existing pathways for tigers to travel between reserves, as more land is developed for other uses such as agriculture or transportation infrastructure. Alternatively, if these corridors are effectively protected, and in some cases expanded, tigers will have a much better chance of persisting.

Effectively protecting tigers in Madhya Pradesh provides indirect benefits to a huge number of species. Many mammals, such as Asian elephants, nilgai antelope, wolves, leopards, and sloth bears, also live in tiger reserves, where they benefit from the protected habitat, reduced poaching, and opportunities to use wildlife corridors between parks. Furthermore, the benefits of connected parks extend well beyond mammals to provide a place for hundreds of species to thrive, including birds, reptiles, insects, and plants.

I take a lot of solace in the fact that efforts to establish networks of connected parks are becoming increasingly commonplace in conservation. This practice extends well beyond tigers in India and includes a growing list of networks designed to protect charismatic species such as brown bears and wolves in North America, jaguar in Central and South America, and elephants in Africa.

Although protecting and connecting parks is currently a successful strategy for many species to thrive, it is not a panacea. First, as was the case with Panna Tiger Reserve, the parks have to be well protected from threats such as poaching and land development. Furthermore, some species are facing challenges that parks alone can't fix. Fortunately for them, nature has some more tricks up its sleeve to rescue life, and people are finding new and powerful ways to give nature a boost.

THE FUTURE OF WILD TIGERS

The connected networks of parks for tigers exemplify the rescue effect in action: when a given population of tigers in the network is stressed or begins to decline, a suite of biological processes can begin automatically, helping the population adjust and recover. In this case, the rescue effect would arise from the interaction of three processes: high birth and survival rates causing uncrowded populations to grow (reproductive rescue), and immigration of new tigers from nearby populations bringing in an influx of new individuals (demographic rescue) and genetic diversity (genetic rescue). Panna Lal's story demonstrates all three of these processes: the product of a tiger baby boom in Panna left his home to inhabit and breed in a new park.

Whether or not the rescue effect will be strong enough for tigers to persist in the future will depend on human behavior. If people value parks and their wildlife, they are much more likely to abide by the rules and make sure tigers are protected. Former Panna field director Murthy thinks that this kind of buy-in was key to the successful tiger recovery in Panna, describing it in Hindi thusly: *"Jan samarthan se baagh samrakshan"* (tiger conservation with the people's support).

The easiest way to build support for wildlife protection is via economic incentives: if people can make a living resulting from the existence of parks, they will be far more likely to help protect these reserves. As nature tourism grows in India, the incentives to protect tigers are increasing. I saw this on my visit to Bandhavgarh in 2017, where locals are employed to work in the park and in the surrounding tourism economy.

Maintaining local support for tiger conservation will also require managing human/tiger interactions. As tigers become more densely packed in parks and seek new territories elsewhere, they will be more likely to stray into human-dominated areas. The Indian government is already working to address the problem of tigers killing livestock by offering owners some compensation for their losses. Tigers may also harm people, however, which risks swiftly eroding support for tiger protection. Maintaining peace between tigers and people will therefore require highly proactive efforts to capture or even cull tigers that move into human-dominated areas.

For now, Panna Tiger Reserve appears to have arrived at a relatively stable state, where local interests and management actions favor tiger conservation. But there is no inherent reason to believe that the current state will remain indefinitely. For example, even if local groups have stopped poaching and begun to support tiger conservation, others could arrive from elsewhere to profit from killing Panna's tigers. Indeed, there are already troubling signs that poachers are returning to Panna. Ultimately, it will be up to wildlife officials to address any upticks in poaching, but their job will be much easier if local people share the same goal. A similar dynamic is likely to play out around all of India's tiger reserves, whereby a combination of effective enforcement and local support will prove most successful in protecting tigers and other wildlife. If either of these elements breaks down, wildlife will be at heightened risk. Ultimately, the combination of nature's rescue effect and the willingness of people to set aside well-protected land will make the difference for tigers as they adjust to a changing world.

Encouragingly, the rising tiger population in Panna Tiger Reserve is not an outlier within India. After India's tiger population bottomed out at about 1411 tigers in 2006, tiger populations have more than doubled to an estimated 2967 animals in 2018. That's an estimated increase of 1556 tigers in twelve years—probably more than all the remaining wild tigers outside of India.

THE REIGN OF PANNA LAL

On our 2017 safari adventure, we stayed with Panna Lal for at least a half mile, backing up the jeep as he continued to approach us. He occasionally paused to mark his territory, and he even took a short dip in an early monsoon mud puddle as we watched. Eventually, he turned off the road and vanished into the jungle, his stripes quickly obscuring his silhouette.

We waited for a while in the jeep to see if he might reemerge. The adrenaline from our close encounter was starting to subside, and as the sun was setting, the jungle seemed unnaturally still.

Then, from not far off came a visceral sound that sent shivers up my spine. I had never heard anything like this before, but there was no mistaking what it meant. The son of Panna was roaring, putting every creature within earshot on notice that he had claimed this territory and was ready to fight to protect it. Tigers have been the rulers of the jungle for thousands of years, and if people are ready to fight alongside them, the reign of Panna Lal and his kin doesn't have to end anytime soon.

CHAPTER 2

An Irresistible Urge

"HEY, BEAR!" MY OLDEST SON, QUINN, shouted to no one in particular. Ten seconds later, I followed with a "Hey, bear!" of my own. It was midsummer of 2014 as we followed tiny Hansen Creek in southwest Alaska upstream, toward a remarkable natural spectacle. Scientists started monitoring this creek in the 1960s, dutifully recording how many salmon returned each year to spawn, and this year shattered the record for the most salmon ever.

Along the way, we pushed through dense brush, sloshed through water, and swatted at mosquitos. Our mobility was limited by our chest waders, gray and tan waterproof pants that were held up with black suspenders. As we slowly made our way along the creek, the water constantly erupted with fleeing salmon that mistook us for predators aiming to attack them from above.

There were no bears in sight. Our regular calls of "Hey, bear!" were in accordance with the standard safety protocol for walking along salmon streams in southwest Alaska. Brown bears in this area grow to unusually large sizes by gorging on the smorgasbord of salmon sashimi that returns to creeks like Hansen every summer. These huge bears are intimidating to encounter at close range, and although they usually cause people no harm, it's a good idea to keep your distance. The best way to avoid a face-to-face confrontation along a densely vegetated creek, where visibility is low, is by announcing to the bears that you're coming.

Sows, female bears, are especially dangerous, because they often have cubs. Sows will defend their cubs from all comers and have been known to charge, and even maul, people who get too close. By announcing our presence with our shouts, any sow upstream would have a chance to move her cubs to safety before we arrived. The same goes for large boars, the male bears that will also usually move off the stream into the bushes when they hear people coming. A few years earlier, on a nearby stream, I had clambered over a fallen tree onto a gravel bar to find a wet, flopping salmon bleeding from a bite mark on its back. Next to the salmon was a trail of water drops leading into the bushes from a bear that had fled just moments earlier. That was proof enough for me that shouting "Hey, bear!" works.

As we made our way up Hansen Creek, Quinn and I paused for a minute to watch the sockeye salmon in the shallow water. We could see a spawning pair hovering over a depression in the gravel, which served as a nest. Suddenly, the female, recognizable by her shorter jaw and more slender body, started quivering against the gravely bottom, releasing some pea-sized red eggs. At the same moment, the male swam alongside her body and released a cloud of white sperm. A second male was watching nearby and tried to get in on the action by swimming quickly into the nest and releasing his own cloud of sperm. Enraged by this threat to his paternity, the first male snapped at his rival's tail with his long, hooked jaw.

When the salmon finished spawning, Quinn and I continued upstream. We were invited here by Professor Thomas Quinn of the University of Washington's School of Aquatic and Fishery Sciences and his colleagues from the university. I had been partnering with these researchers for many years to understand the ecology of sockeye salmon in the Bristol Bay area of southwest Alaska. When we arrived, Professor Quinn (not to be confused with my son Quinn) told us excitedly that although the salmon run had already passed its peak, if we hurried to the stream we could see the tail end of an extraordinary run.

Most of the remaining salmon were now at the head of the creek, where an artesian spring helped feed a large pool of cold, crystal clear water, and a waterfall blocked the salmon from traveling farther upstream. As Quinn and I finally reached the pool, I looked into the water and was gobsmacked. When I was a graduate student, I lived in the Pacific Northwest state of Oregon, where I heard many tales about salmon runs "in the good old days." Old timers would say that there used to be so many salmon that a person could walk across a stream on their backs without getting wet. I always took these stories to be fanciful exaggerations, because I never saw more than a few salmon swimming in Oregon rivers at that time. However, in the pool at the head of Hansen Creek, it was unmistakably "the good old days" for sockeye.

We walked into the pool as thousands of salmon swirled around our feet. Quinn was fascinated by the fact that he could just reach down near his ankles and pick up fish. Like all Pacific salmon, sockeye die shortly after spawning. Most of these had already spawned, and with their duty to the next generation complete, they were just waiting out the clock on the inevitable.

Although I found the scene spectacular, the extraordinary annual return of sockeye salmon to Hansen Creek in 2014 was just one part of a much bigger pattern. Thousands of small populations of sockeye, similar to the Hansen Creek population, are spread across Bristol Bay. The long-term record shows that the number of spawners in each population bounces up and down from year to year as a result of chance events. Collectively, however, these individual populations have given rise to a resilient ecosystem that today is producing more salmon than ever, even after a century of intensive fisheries and decades of climate change. The key to Bristol Bay's resilience is the rescue effect, and especially reproductive rescue, which fuels its productive fisheries.

THERE AND BACK AGAIN

Most of the salmon in Hansen Creek in 2014 had started their lives in this same gravel four years earlier. Following their parents' spawning efforts, the embryos spent the fall developing in the spaces between the rocks underwater, before hatching in the late fall. After hatching, they continued to grow and develop in the gravel before emerging into the stream in the spring and following it to nearby Lake Aleknagik. Here, they plied the lake's cool, clear waters, hunting and eating food such as tiny zooplankton adrift in the water, while trying to avoid being eaten themselves by larger fish or by birds.

After a year in the lake, the young sockeye headed downstream again, this time all the way to the Pacific Ocean—in general, there's a lot more food in the ocean than in lakes and streams. Once in the ocean, the silvery sockeye started a two-year growth spurt on a diet of zooplankton and larger foods such as squid and fish. This growth spurt came to an end in the early summer of 2014, when the fish developed an irresistible urge to complete their lifecycles by returning to Hansen Creek to spawn. They then began an epic journey across hundreds of miles of ocean and through a labyrinth of rivers and lakes. To find their way, the salmon used a sense that humans don't possess, which works like an internal compass that measures the Earth's magnetic field. They also noted minute differences in the scents from different waters that came from different rivers and lakes to smell their way home, retracing the same route they took downstream years earlier.

About a month before spawning, the sockeye started an extraordinary metamorphosis. At first, their silver skin developed the slightest red blush, like a ripening apple. As they neared Hansen Creek, the process accelerated, and their blush intensified to crimson, their heads turned green, and, especially for the males, the tissues on their backs swelled into pronounced humps and their jaws elongated into tooth-studded hooks. The salmon also stopped eating and started drawing on the precious fat and protein stored in their bodies to fuel their migration and grow supersized gonads: ovaries with thousands of eggs for the females and large testes for the males. All these changes were in service of the evolutionary purpose of their lives: to achieve immortality by successfully producing a next generation.

It was not until the sockeye reached the mouth of Hansen Creek that anyone realized that something unusual was going on here in 2014. Before entering the creek, the salmon had collected in schools along the lakeshore of Lake Aleknagik. It's difficult to count salmon when they are staging like this, but you can get a rough idea how many are returning by estimating the size of the school from above. Professor Quinn has been studying salmon for more than four decades, and he has taken a special interest in Hansen Creek, returning year after year to study its salmon. Before I'd arrived, he had been visiting the area daily, and he told me that the reflection of the blue sky in the lake, combined with the red color of the salmon, made a huge part of the lake look purple. It seemed like the creek might be in for a big run.

To reach their spawning grounds, the Hansen sockeye had to leave the deep water of Lake Aleknagik and swim upstream through Hansen Creek's shallow water. In 2014, Professor Quinn likened the scene at the mouth of the creek to an escalator, as he watched scores of salmon pass by but couldn't see those that were still coming up from the lake. That year, the salmon escalator at the mouth of Hansen Creek was jam-packed with thousands of salmon coming up, day after day. At the peak of that spawning season, he counted 5000 salmon entering the stream every single day. Those daily counts comprised about as many salmon as returned in a typical year since scientists from the University of Washington started measuring sockeye returns to this creek in the 1960s. Previously, the largest number of salmon recorded at the creek in a single year occurred in 1966, with about 19,000 fish. This year's spawning season easily shattered this record, as more than 55,000 sockeye salmon—ten times the average and three times the previous record—crammed into this tiny stream to spawn.

THINNING THE RANKS

People have been keenly interested in Bristol Bay's salmon as a bountiful and relatively easy-to-catch source of food for thousands of years. During most of this time, fisheries occurred at a small scale, as Alaskan Natives caught food for local consumption. These Alaskans were naturally constrained in their catch, because the salmon spoiled quickly, and the technology used for preserving it, such as smoking and drying, was labor intensive. Furthermore, Bristol Bay's salmon were so abundant that these fisheries probably didn't have much effect on the overall numbers of fish. However, things fundamentally changed with the advent of fish canneries in the nineteenth century, which could quickly process enormous amounts of fish into shelf-stable cans for shipment to distant markets.

Bristol Bay's first cannery started operations in 1893, when it processed almost a million salmon. As more canneries were built and more boats arrived to catch salmon, the commercial fishery ramped up quickly, catching ten million fish a year by 1901. Since then, the commercial catch has averaged more than fifteen million salmon per year, with exceptional years reaching forty million.

The commercial fishery intercepts the salmon as they journey from the ocean to their freshwater spawning grounds, with most fish caught near the mouths of Bristol Bay's five major rivers. In June of every year, the captains of about 1300 boats descend upon Bristol Bay, readying their crews and gear. By mid-June, the salmon start pouring into the rivers, and fishery managers at the Alaska Department of Fish and Game give the much-anticipated greenlight for the boats to deploy their curtainlike gill nets. Each boat unreels its transparent net in front of the salmon—and near many other boats—in a choreographed race to catch as many fish as possible. If salmon are in the vicinity, they immediately start colliding with the net, which quivers as the fish become entangled by their gills. As the net is reeled back in, the doomed salmon are untangled by hand and stored in the boat's hold. Fishing boats repeat this process over and over for as long a period as they are allowed, or until their holds are so full of salmon that they run out of room to store any more. (Professor Daniel Schindler of the University of Washington School of Aquatic and Fishery Sciences told me that the catches in 2020 and 2021 had been so high that some boats actually sunk from the weight of taking on too many salmon at one time.) Periodically, fisheries managers will halt the fishing in accordance with the sacred maxim of Alaska salmon management: let some fish escape upriver to spawn.

The rapid growth of the commercial salmon fishery more than a century ago represented a major change to the salmon's ecosystem. Left to their own devices, most salmon populations will eventually arrive at a carrying capacity, in which the number of offspring entering a population is about the same as the number of individuals that die, so that the population size naturally oscillates around an average value. With the rise of commercial fisheries in Bristol Bay, sockeye salmon were pushed well below their carrying capacity as they ran a deadly gauntlet of hundreds of miles of fishing nets.

By thinning the ranks, the fishery has become a boon to the fish that manage to escape. When the salmon arrive at their spawning grounds, they are typically less crowded than they would be without the fishery. With less competition, the fish can spawn in the best sites, where their fertilized eggs will be most likely to survive, and later spawning salmon are less likely to dig up existing nests, sending their eggs downstream into the hungry jaws of waiting trout. When the tiny salmon emerge the next spring, they find the lakes less crowded than they otherwise might be, giving them a chance to find more food and grow more quickly.

By holding the salmon below their carrying capacity, the fishery is purposely triggering reproductive rescue in the form of a baby boom. During a baby boom, large numbers of salmon survive to adulthood, so populations increase. In the part of Bristol Bay where Hansen Creek is located, a typical spawning fish today contributes not just one offspring to replace itself in the next generation, but three. If the fishery were restricted, and all of those fish were able to return to spawn, the fish density in the streams would triple, which would increase the risk of overcrowding, and future returns may decline as a result. Instead of allowing all the salmon to return, fishery managers aim to remove about two-thirds of them prior to spawning. In doing so, the fishery managers deliberately keep driving the spawning population back below the carrying capacity, making for good spawning conditions and high survival rates for the next generation.

By turning the fishery on and off and counting the fish that squeeze past, fishery managers aim to titrate a perfect balance: catch as many fish as possible, while letting just enough escape to produce another strong return. Although managing for reproductive rescue doesn't always work in fisheries, Bristol Bay sockeye have been successfully managed this way for more than a century. Since then, reproductive rescue has enabled the Bristol Bay fishery to sustainably harvest more than two billion salmon.

Reproductive rescue works not only for managing salmon; it sits at the heart of many kinds of harvest management, whether for tuna, trees, or turkeys. Most populations, if pushed below their carrying capacity, will automatically begin to rise through reproductive rescue, making it probably the most robust and ubiquitous way that the rescue effect operates in nature.

A BALANCED PORTFOLIO

The salmon that return to Bristol Bay every summer come from hundreds of smaller populations, such as the one in Hansen Creek. Those that make it past the fishing nets fan out, taking separate paths up rivers and through lakes in search of the spawning sites where they hatched years earlier. Some of these spawning sites are in big rivers, others are in lakes, and some, like Hansen, are in small creeks. Each spawning site varies in other ways too, including water temperature, water flow, and nearby food availability for young salmon, which can affect the ideal time to spawn, hatch, and migrate downstream. Through thousands of years of evolution by natural selection, different spawning sites have given rise to hundreds of local populations that have been molded to fit their surroundings like a glove.

My favorite example of this local adaptation is demonstrated in dramatic fashion in the appearance of the spawning salmon. Sockeye choose mates in part based on their physical appearance. To a spawning sockeye, for example, a disk-shaped body can be considered very sexy. Lake-spawning sockeye, especially the males, have evolved to grow enormous hump-shaped backs to impress females at the spawning grounds. Lake-spawning males have access to deep water for spawning, and they benefit from their attractiveness to females. In small creeks, however, where salmon may spawn in water as shallow as 4 inches (10 cm), a high hump is a liability—for example, a bear can easily catch a salmon whose back is conveniently sticking out of the water. Creek-spawning sockeye have smaller humps that are more likely to stay submerged in shallow water, making it easier for them to sneak past the jaws and claws of predators.

If a sockeye were to spawn in the wrong place, it may find itself mismatched—perhaps spawning at the wrong time or having a body shape that makes it a target for hungry bears. In addition, there's a decent chance that any offspring it manages to produce will be less successful. To improve their chances of reproductive success, sockeye are instinctively drawn to perform their last biological act in exactly the same place that they performed their first. After all, if it worked for their parents, it will probably work for them.

As a result of this strong homing instinct, each salmon population exists in a particular place, with locally adapted fish, making each population somewhat independent from the next. When one population experiences especially good conditions, it will rise above neighboring populations; with poor conditions, it will drop below. This is probably what happened on Hansen Creek in 2014, when, for some unknown reason, this population experienced off-the-charts survival rates. Maybe there were unusually good water conditions for development in the gravel of Hansen Creek. Maybe they timed their entry into Lake Aleknagik perfectly and found an abundance of food. Their chance good fortune came with a price, however, as crowded spawning conditions may have led to poorer success for the class of 2018, which was roughly half the size of the 2014 run.

Over time, the salmon return data at Hansen Creek looks like a roller coaster that periodically rises and falls. At the same time, other populations are following their own roller coasters of abundance. Since no two local populations are experiencing the exact same conditions, their populations tend to be out of sync; when one is high, another might be low, and vice versa. With so many populations experiencing their own ups and downs, the whole system operates like a diversified stock portfolio. By spreading investment risk across dozens of financial stocks in different markets, which have their own ups and downs, an investor can be protected against the all-eggs-in-one-basket risk of holding only one kind of stock. Similarly, the overall returns of salmon to Bristol Bay are far more reliable because they comprise fish from so many individual populations, with each growing and declining on its own schedule.

Reliability is important to fishers, who invest heavily to buy permits and boats, hire crews, and incur other costs. During bad years, when few salmon return, fishers bear the costs with little to no gain to offset them. Fortunately, the diverse portfolio of sockeye populations means that bad years are rare. Since its inception, the fishery has never had to shut down entirely as a result of scant returns, and since 1980, the annual catch of sockeye has dropped below ten million only once.

The balanced portfolio of sockeye populations offers real benefits for the ecosystem too. Predators such as brown bears, bald eagles, and rainbow trout depend on the annual salmon return for food. If the salmon are faring badly in one area, these mobile hunters can walk, fly, or swim to a place where they are more abundant. One of the reasons why Quinn and I were so diligent in shouting "Hey, bear!" along Hansen Creek is because the predators in the area knew about the big salmon returns, and they, like us, had descended on the creek. Though we were happy to see eagles and trout, we were hoping to avoid the bears.

Ultimately, the portfolio of out-of-sync salmon populations creates strong benefits for the salmon populations themselves. That's because, in addition to rescuing their own population through reproduction, sockeye salmon populations also rescue one another—by making some mistakes on their journeys home.

STRATEGIC MISTAKES

The homing ability of a sockeye salmon is impressive, but not perfect. The Bristol Bay salmon experts that I spoke with estimated that at least 99 percent of Bristol Bay's spawning salmon make it back to the places they hatched years earlier, although it's difficult to measure this exactly. That means that about 1 percent of the returning salmon spawn some-place new; in the salmon world, these lost souls are known as *strays*. Although the overall rate of straying is low in sockeye, it still adds up to a lot of individuals. For example, in 2020, 18.7 million sockeye eluded the fishery to spawn in Bristol Bay. If 1 percent of the population that year strayed, that would mean that 187,000 fish spawned somewhere other than home.

For an individual salmon, straying may be a bad idea, because it could wind up spawning in a place where it or its offspring are ill suited to survival. Worse yet, if a stray is terribly lost, it may not find a place to spawn at all. If this is the case, then straying is best explained as an unfortunate mistake that arises from an imperfect system for finding home. But because there are several potential benefits of straying, some salmon biologists see straying as a strategy, and not a mistake.

To see how straying could be advantageous, let's imagine that Hansen Creek came upon hard times, and its population dwindled to a handful of adult fish returning to spawn. Such a small population could easily go extinct from just one hungry bear looking for breakfast. But if a few extra fish strayed into Hansen Creek, their numbers might be the difference between the population persisting and disappearing. In this case, Hansen Creek's sockeye run would be experiencing demographic rescue, similar to what occurred when baby corals drifted in to rescue Fiji's cyclone-ravaged reefs. Both cases show that immigrants from another population can rescue a small population by boosting its total population size and its future reproductive potential. Demographic rescue makes it nearly impossible for any sockeye population to disappear in Bristol Bay, because there are always some strays in the mix to give it a boost.

For the population on the receiving end, straying salmon can be a double-edged sword, genetically speaking. The outsiders invariably bring some new genetic diversity, so if strays are able to spawn successfully, the next generation of salmon will inherit some of the strays' traits. If the local population is already well adapted to its environment, and the strays are poorly adapted, incoming strays can dilute the local gene pool via outbreeding depression. Imagine, for example, that a lake-spawning sockeye passes on its genes for a large, humped back to a population of shallow-creek spawners; this would make its offspring easy pickings for bears when they return to the area to spawn.

Alternatively, an influx of new genetic diversity might be exactly what a small population needs. If a sockeye population were to fall to just a few fish, the chances would increase that every spawning fish would be closely related, perhaps siblings or cousins. Just as with tigers isolated in reserves, mating between closely related salmon could cause inbreeding depression, which involves a loss of genetic diversity and higher risk of genetic disorders. For such a population, strays could bring more than just new individuals to the population (demographic rescue), but also new genes. When mating occurs between the prior residents and the newly arrived strays, the whole population's genetic diversity can increase. In doing so, the population can overcome the genetic disorders that are common in small populations—this is genetic rescue.

Occasionally, strays hit the jackpot by finding vacant habitat where they can establish an entirely new population. For example, straying could happen in the ocean, resulting in some salmon entering the wrong river. If that river doesn't already have a salmon population of its own, a handful of strays might be able to establish a new population. And, in fact, this is probably how sockeye arrived in Bristol Bay around 15,000 years ago.

During the last ice age, which peaked around 25,000 years ago, Bristol Bay was much colder and largely buried in glaciers, making it difficult or impossible for sockeye to survive there. However, sockeye salmon were able to find refuge elsewhere, perhaps in rivers and lakes on Kodiak Island in the Gulf of Alaska and farther south in the Columbia River that flows into the Pacific along the border of Oregon and Washington. When that ice age ended, the glaciers began to melt, and water flowed over the newly exposed land, forming rivers and lakes. Just by chance, some strays eventually reached Bristol Bay and established populations, which persist today.

With human-caused climate warming well underway, there are already signs that salmon are on the move again. Professor Daniel Schindler told me in 2022 that the Kuskokwim River, north of Bristol Bay, has seen record returns of sockeye since 2017, with unusually warm conditions in the far north. He said that Kuskokwim residents are catching more sockeye than anyone can remember. Farther north still, in rivers flowing into the Arctic Ocean, are many accounts of changes to salmon runs, with some longstanding runs growing and new runs appearing. Though runs are still small in these Arctic rivers, they may accommodate major salmon populations in a warmer future.

When populations shift from one place to another in response to environmental change, they are being rescued by their ability to colonize new geographic locations. This is geographic rescue, which occurs when some individuals disperse to new locations and by chance find favorable environmental conditions. With the world's climate rapidly changing, geographic rescue is becoming key to the survival of many species as they diminish in places where they have historically thrived and begin to thrive somewhere new.

According to Professor Quinn, homing and straying are the "yin and yang of salmon." Too much straying dilutes the local gene pool, undermining adaptation. But perfect homing would be a dead end, because it would prevent movement from one place to another and could result in a shallow gene pool. If salmon stray just a little, populations can get a numerical boost from immigrants (demographic rescue), a periodic influx of new genes (genetic rescue), and the ability to relocate if need be (geographic rescue). By having a balance of homing and straying, salmon species have the best of both worlds: populations that are well adapted to local conditions and the ability to adjust quickly if conditions change.

FIGHTING TO SURVIVE

Bristol Bay sockeye show the rescue effect in action: whenever one population heads into decline, reproductive, demographic, and genetic rescue can automatically bring it back. The rescue effect has made the Bristol Bay sockeye so resilient that they have withstood more than a century of humans catching about half of the population annually, while simultaneously adjusting to a changing climate. And if climate change ever renders Bristol Bay unsuitable for sockeye, then geographic rescue, fueled by straying salmon, may help them relocate elsewhere.

Bristol Bay sockeye are not unique in benefiting from the rescue effect; the same processes that enable them to thrive as their world changes are at work for many species. Most populations experience reproductive rescue when their numbers become low, which enables them to rebound. Populations are often connected to each other through immigration, fueling demographic and genetic rescue. Moreover, any species whose dispersal sometimes takes them to new places—such as salmon that stray, seeds that are blown by the wind, or animals that migrate—have a shot at undergoing geographic rescue.

Sockeye salmon experience an irresistible urge to swim across the ocean and then travel upstream, all the while dodging predators such as humans, bears, and eagles, only to fight one another, tooth-and-jaw, for a singular chance to seed the next generation. Although the details of their story may differ from those of millions of other species on Earth, the underlying plotline is the same: individuals will always fight to survive, putting everything they have into creating a successful future for their offspring. In the decades to come, most species will persist because of this drive to survive and procreate. For all the rest, people will get to decide whether and how to give them a boost.

CHAPTER 3

Playing Possum

THE PYGMY-POSSUM DUCKED INTO a crevice in the rocks. She had heard something in the moonless night and was patiently waiting for any danger to pass. This little creature, with soft, gray-brown fur, big black eyes, and a pointed snout, resembled a mouse. Unlike a mouse, however, she had a long, prehensile tail, which she was currently using to carry bundles of moss to add to the bedding in her nearby nest.

In spring, she had given birth to four tiny, pink, helpless babies. After birth, each baby crawled into her pouch, attached to a teat, and began nursing. A few weeks later, they were out of the pouch and exploring the nest. Then, in early summer, her young left home to fend for themselves. But raising them had cost her a lot of energy and she was out late on this night looking far and wide for food to build up her fat reserves to prepare for the coming winter.

The pygmy-possum lived in an area of hilly forests with abundant rock piles that provided her with cover from predators, places to build her nest, and plenty of insects, nuts, and berries to eat. She had survived here for four years and had raised three litters of offspring. Her children, grandchildren, and even great-grandchildren lived in the landscape around her.

After she decided that she had waited long enough in the crevice for any danger to pass, she started out across the forest floor toward home. This was a fatal mistake, for in a nearby tree, a masked owl listened intently to her footsteps on the leaf litter. Concluding that his prey was now out in the open, he dropped from his branch and glided silently toward the pygmy-possum. She didn't hear him coming.

With a flash of his talons, the owl pinned the pygmy-possum to the ground, swiftly constricting the life from her tiny body. When she stopped struggling, the owl grabbed her head with his beak, and, lifting her up above his head, he swallowed her in one great gulp. Satisfied with his meal, the owl flew to his roost in a nearby cave. Soon it would be first light and he'd rest for the day, contentedly digesting his night's catch.

The next day, just before nightfall, the owl regurgitated a pellet. He could digest only the soft parts of his prey, so he coughed up the bundle of indigestible fur and bones from his meal the previous night and let it fall to the cave floor.

Thousands of years later, in the late nineteenth century, paleontologist Robert Broom found the remains of the owl's cave, and, once inside, he found traces of its former inhabitant's meal in the form of a fossilized owl pellet. Over the years, the fur in the pellet had decomposed, but some of the bones had fossilized and fused together in limestone that formed in the cave. As Broom carefully chipped the limestone away and separated the individual bones inside, he found the fossilized jawbone of the pygmy-possum.

Broom noticed that this jaw was different from any he had ever seen, especially its premolar tooth. Our human premolar teeth sit between our canines (the sharp, pointy teeth near the front of our mouths) and our molars (our back teeth). The premolar of the ancient pygmy-possum's jaw jutted up above the nearby teeth, like a rounded shark fin rising above the water. No known living species of possum possessed such a tooth, nor did any species that had been described in the fossil record. Because Broom recognized a resemblance between this tooth and one from a rat-kangaroo, he concluded that the two animals might be related.

In 1896, Broom published a scientific paper describing this jawbone as well as others that he had found. He concluded that the jawbone was from a previously unknown species of small marsupial, which he named *Burramys parvus* (small rock mouse). For the next seventy years, scientists debated where *B. parvus* fit into the tree of life. With improved tools and more bones, they concluded that it was not a close relative of rat-kangaroos after all, but was instead a species of possum.

Possums are native only to Australia, New Guinea, and nearby islands, and these animals are different from the opossums that I knew growing up in the United States. Both are marsupial mammals, as are kangaroos and koalas, but the possum and opossum lines diverged roughly eighty million years ago, well before the evolutionary split between dogs and cats, as Australia drifted away from the remnants of Gondwana to become an isolated island continent.

Today, about seventy known species of possums have diversified to fill many different ecological roles. Some are ecologically similar to the opossum, and these raccoon-sized mammals feed on both plants and animals. Others have evolved to resemble flying squirrels, complete with winglike skin that helps them glide from tree to tree. The pygmy-possums are the smallest of these and include a handful of living species that look similar to mice or hamsters. The fossilized *B. parvus* was eventually identified as a pygmy-possum, but it was quite different from any of the living pygmy-possums that were known from Australia and New Guinea.

The story of *B. parvus* could have easily ended as little more than a scientific footnote—just another extinct animal that was identified through fossils and cataloged for scientific posterity. The Earth has lost untold millions of species to extinction, after all, and when Broom found the fossilized bones in the cave, there was no reason to think that the little animal warranted further attention. As it turns out, however, this was not the end of the story.

AN UNEXPECTED FIND

In the interior of the Australian states of Victoria and New South Wales are the Australian Alps. Individual peaks in this range rise more than 6000 feet (1830 m) high, which is quite high by Australian standards. In the winter, the Australian Alps see enough snow to support Australia's only downhill ski resorts.

In 1966, Professor Ken Shortman, an immunologist from the Walter and Eliza Hall Institute of Medical Research in Melbourne, decided to go on a weekend ski holiday in the Australian Alps with friends from a University of Melbourne ski club. Their destination was a university-owned a ski hut high up on Mount Hotham in Victoria. In those days, there was no passable road to the mountain in winter, so Shortman and his colleagues parked at a lower elevation, packed their rucksacks and skis, and hiked up to the hut.

During their stay, one of the skiers caught a curious creature searching for crumbs in the kitchen. Some of the skiers thought it might be a rat, suggesting that perhaps someone should set some traps in the hut.

The creature turned out to be quite tame. When I spoke with Shortman more than fifty years later, he recalled that he "befriended it, and it sat on my shoulder and we fed it honey and cheese."

As the skiers bonded with their unexpected hutmate, which they named George, they argued about what kind of animal it was. One skier thought it a phascogale—a mouselike marsupial that inhabited other parts of Australia. Someone else was sure it was a kind of possum, because of the shape of its paws. But they all agreed on one thing: it didn't belong at this altitude on this snowy mountain. According to Shortman, "No Australian creature would live above the snow line in the Alps, so we figured that it must have got into the hut on logs of wood that are sent up by truck in summertime to stock the ski hut."

Still puzzled about the creature's identity, Shortman decided to take it home to Melbourne, where he would try to identify it. At the end of his ski holiday, he carefully packed George into a large biscuit tin. After loading the tin in his rucksack, Shortman ventured out in the middle of a blizzard and skied down the mountain carrying his new friend.

Promptly upon returning home, Shortman went to the University of Melbourne to try to identify the creature using charts of Australian mammals. But it wasn't there.

Still determined to solve this mystery, he once again packed the little creature into the biscuit tin and took it to the experts at the Fisheries and Wildlife Department of Victoria, where he showed it to Senior Research Officer Robert Warneke, expecting him to know immediately what he had found. Instead, Shortman said the first two comments from Warneke were, "God, what is that?" followed quickly by, "Look at the balls on it!" Apparently, George was a male.

When Warneke examined the creature, he initially noted that it strongly resembled the long-tailed pygmy-possum (*Cercartetus caudatus*), known only from northeastern Queensland. But he wondered,

"What was a long-tailed pygmy-possum doing in Victoria?" After a closer inspection, he quickly determined that this clearly was not the same animal.

Warneke recognized that the animal must be an undiscovered species of pygmy-possum. Under Victorian law, all indigenous fauna, regardless of taxonomic status, was automatically protected, and Warneke "had a clear responsibility to secure the animal for identification and scientific study." So he asked Shortman to leave the pygmy-possum in his care. Shortman was initially reluctant, because he wanted to give George to his four-year-old daughter as a pet. Warneke then explained that it was inappropriate to keep Australian mammals as pets, after which Shortman relented and turned over his find. In compensation, Warneke gave Shortman a consolation prize for his daughter: an eastern pygmy-possum (*Cercartetus nanus*), a common species in Victoria that was being kept in the laboratory for observation and breeding. Warneke admitted to me that it "was quite illegal for me to do that."

Warneke put the mysterious pygmy-possum into a locked cage in his laboratory for safekeeping. He then reached out to a colleague, Norman Wakefield, with whom he had published taxonomic studies on other marsupials, knowing he would also be intrigued by the discovery of a new species. A few days later, Wakefield showed up at the laboratory, unannounced, while Warneke happened to be absent. Wakefield was so eager to examine the pygmy-possum that, rather than waiting for Warneke to return, he "cajoled" Warneke's laboratory assistant into unlocking the cage. Then, using the handle of a plastic spoon, Wakefield gently pried the creature's mouth open. And there it was, the telltale premolar, the likes of which was known only from the fossil record. In fact, it was an exact match to the premolar described by Robert Broom in 1896. It turned out that *B. parvus* was playing an evolutionary version of possum—appearing to be extinct, while actually being alive and well.

THE MOUNTAIN PYGMY-POSSUM

Since the discovery of a living *B. parvus*, the animal has been given a common name: the mountain pygmy-possum. Contrary to the skiers' conclusion, it does indeed live above the snowline on Mount Hotham. In fact, one colony lives in and around the ski hut, and Shortman encountered the little animals for years when he returned to ski.

Later explorations have discovered more mountain pygmy-possums on nearby mountains and high-elevation plains. Two populations were also found farther away: one to the northeast on Mount Kosciuszko in New South Wales, and another to the southwest on Mount Buller in Victoria. All three populations live in some of Australia's highest mountains at elevations ranging from about 4000 feet (1220 m) to the very top of the Australian continent's highest mountain, Mount Kosciuszko, at 7310 feet (2230 m).

In the half-century since the first living mountain pygmy-possum was discovered, scientists have learned a lot about how the little marsupials survive in extreme conditions. And they have also started sounding the alarm, because these furry creatures are now truly barreling toward extinction. Although there was no lament for its apparent extinction thousands of years ago, now that we realize it actually survived, should we feel responsible for making sure that it doesn't disappear on our watch? And if we take that responsibility, how far are we willing to go to try to ensure that these possums persist?

A CLIMATE CHANGE RELIC

No one knows for sure how mountain pygmy-possums ended up living on the high peaks of the Australian Alps, but the answer almost certainly has to do with climate change.

Professor Mike Archer, a paleontologist at the University of New South Wales, describes himself as "an outrageous, irreverent, and hopefully innovative conservationist." He's spent years studying extinct Australian mammals, and he is devoted to the effort to save the mountain pygmy-possum from extinction. After studying the fossil record, he noted that the possum's ancestors had not, in fact, thrived in the alpine; instead, they lived in lowland rainforests for millions of years.

The fossilized pygmy-possum jawbone found by Broom belonged to an animal that probably lived during the Pleistocene, the last geological epoch, which included the last ice age. The climate in the Pleistocene was not always cold, however; instead, the Earth at that time was cycling through periods of warming and cooling, which were causing glaciers in the Northern Hemisphere to retreat and advance accordingly. In the Pleistocene, Australian glaciers were not widespread, but the climate cycled through cooler, drier glacial weather followed by warmer, wetter interglacial weather. The lower dwelling ancestors of mountain pygmy-possums likely flourished during the interglacial weather, as the more favorable wet forests expanded. Archer believes that "during one of the peak wet periods in an interglacial [period], those rainforests went sweeping up the flanks of the mountains to the alpine zones and with them went the possums."

When the climate shifted back to glacial weather, Archer said, the rainforests were "obliterated." The pygmy-possums had lost their preferred habitat, and their geographic range contracted. During one of these wet/dry cycles, conditions presumably became so bad that the last of the lowland pygmy-possums were wiped out. But, somehow, a few pygmy-possums managed to develop an entirely new lifestyle, including hibernating under the snow, which enabled them to persist on the mountain tops.

When the Pleistocene ended roughly 11,700 years ago, the warm, wet lowland forests returned. But apparently the few remaining mountain pygmy-possums never made their way back down to them. Archer believes that they couldn't make it all the way to the lowland forests now: "There's no forests to go through and there's no rock piles to give them protection on the way down to the lowland forests." If they tried to travel to lower elevations, they'd be too exposed to predators and wouldn't find sufficient shelter from the elements. So, today, all the remaining populations are trapped in their high-mountain redoubts.

LIFE IN THE HIGH COUNTRY

Life for a mountain pygmy-possum is a race against time. Each animal is desperately trying to consume and gather enough food during the short summer so that it can reproduce and then survive months in winter hibernation. Although the possum's diet includes large quantities of nuts and berries, one of its most important food sources is the Bogong moth. These large moths are so rich in fat and protein that one researcher described them as "hamburgers on wings." Some describe them as tasting "sweet and walnut-like," and they have long been a food source for Aboriginal Australians.

Australian Bogong moths are known for their biannual migrations, when billions of insects take to the wing in great swarms. They made international news when millions of moths invaded the Sydney Olympics in 2000. *The New York Times* described the scene as follows:

"Millions of them, jammed almost wing-to-wing around the huge stadium lights, appearing to TV cameras like shimmering stars." Another news source reported, "As legendary soprano Yvonne Kenny sang the Olympic Hymn in a dazzling purple gown, a ruffled Bogong moth crash landed on her torso." One swarm was so large that meteorologists mistook it for a rain cloud.

Bogong moths visit the Australian Alps only in the spring and summer; for the rest of the year, they complete their lifecycles at low elevations. They lay their eggs in lowland soils in the fall, which hatch into caterpillars that munch on plants in the relatively cool, wet lowland winters. When a caterpillar reaches sufficient size, it begins its transformation into a moth underground. In springtime, the 1-inch (3 cm) long moths hatch and promptly migrate to the highlands, where they wait out the hot, dry lowland summer before returning to breed in the fall.

By migrating to the mountaintops, Bogong moths subsidize the mountain ecosystem with food from the lowlands. For the mountain pygmy-possum, the Bogong moth migration sets the calendar for their whole year. The arrival of the moths in spring creates a literal windfall of abundant and nutritious food. Possums come out of hibernation just in time to feast on the moths, which helps them recover some of the fat they consumed over the winter and provides females with the nutrition they need to rear litters of young.

And, better still for the possums, the moths spend their dormant summer months crowded together in caves and crevices in the rocks— the preferred possum nesting places. Although it's not clear which came first—the possum nests or the moths—the two are inherently linked: dormant Bogong moths are easy meals at a time when the possums most need abundant food.

By fall, mountain pygmy-possums have completed their reproductive cycles for the year, and their young have matured and left the nest. Each possum has cached some nuts (which it slices open with its remarkable premolar tooth) and berries to help tide it over in the winter. The moths are gone, and winter is coming. Because there is little available food, each possum lowers its metabolism and hibernates. Hibernation can last seven months as they await the return of the moths in the spring. Winter is a dangerous time for these little possums, because if they get too cold or they have insufficient fat reserves, they can die.

If the world were not changing, mountain pygmy-possums would continue to complete this same cycle, year after year, in their mountain homes. Unfortunately for the possums, the world is changing—and change is happening so quickly, that it's not clear whether they will be able to survive.

NO TIME TO MUCK ABOUT

Those who study the mountain pygmy-possum are witnessing troubling declines in their populations. For example, the Mount Buller population in Victoria declined from roughly 350 to 40 individuals between 1996 and 2008 (an 88 percent decrease), and the Mount Kosciuszko populations dropped 44 percent between 1997 and 2009. Reproductive rescue has since kicked in, and populations have recovered. However, the fluctuating numbers of pygmy-possums create a huge risk that populations will wink out of existence during one of the downswings. There are many reasons for these population swings, but they're all related to the species' reaction to changes in their environment.

One of the first changes to affect the possums was a shift in their predators. Before Europeans arrived in Australia, nearly all of the land mammals were marsupials, which had evolved independently from mammals on the other continents for millions of years. When Europeans first began colonizing Australia in the late 1700s, they didn't come alone. They brought a menagerie of new species, including new predators such as foxes and housecats, which have since made permanent homes on Australian soil. Foxes and feral cats have spread far and wide in Australia, even to remote alpine areas where they are more than happy to consume pygmy-possums when they have a chance.

The ski resorts that have been built in mountain pygmy-possum habitat have also taken a toll on the animals. For example, a road built to access a ski resort on Little Mount Higginbotham, near Mount Hotham, cut through pygmy-possum habitat, and it divides the females, which prefer to live at higher elevations above the road, from the males that live below. When the males seek out the females during the spring mating season, they are sometimes killed by cars as they cross over. Even worse, feral cats and foxes wait near the road to intercept pygmy-possums as they cross.

Mountain pygmy-possums are probably most threatened by climate change. Increasing average temperatures around the world have triggered countless biological chain reactions that are altering every ecosystem on the planet. In some cases, threats from these chain reactions are straightforward, such as deadly heat waves. But in many cases, they result in unexpected shifts that are difficult to predict.

No one knows how climate change is impacting the mountain pygmy-possums better than Dr. Linda Broome, who is a senior threatened species officer at Australia's Department of Planning, Industry and Environment. Broome is arguably the world's foremost expert on these marsupials and has studied them for thirty-five years. She and her colleagues have identified at least four ways that climate change is threatening mountain pygmy-possums.

First, climate warming is causing pygmy-possums to freeze to death. This seemingly paradoxical outcome is a result of changes in snowfall. In a normal year, the deep snow that covers pygmy-possum nests insulates them from the coldest winter temperatures, which can drop to a frigid −4°F (−20°C). Deep snow cover helps maintain a temperature of about 36°F (2°C) inside their nests, but when snow is scarce or it melts prematurely, or in wet and rainy winters, colder air seeps in. As their nests grow colder, hibernating possums must burn more body fat to stay warm, and they must wake up periodically to eat some of their precious food reserves to survive until spring. If their reserves run out before spring, they can starve or freeze to death. As the climate in the Australian Alps is warming, winters are less snowy, leaving uninsulated pygmy-possum nests increasingly out in the cold. Archer told me that two such poor snow years in a row "could well cause the collapse of the populations."

In addition, climate change in the lowlands is interfering with the life-cycle of the Bogong moth. Recent years of intense drought have coincided with crashes in moth populations, perhaps because drought-stricken plants were unable to supply Bogong caterpillars with enough food. Facing the loss of their most important food source, pygmy-possums are caught between a rock and a hard place: if the possums travel farther from their rock-pile nests to find more food, they are more vulnerable to predation, but if they instead stay close to home, they won't have enough food to build up their bodies' fat reserves to survive hibernation.

The problem of feral predators—especially cats—may also be worsening as the climate warms. "The cats are probably killed off in years of high snow cover around the alpine areas," Broome explained. "But with lower snow cover . . . we're getting a great influx and higher survival of cats."

Finally, climate change seems to be triggering larger and more frequent wildfires. Although fires were historically rare in the alpine zone, in the last two decades, wildfires have spread into two of the three known mountain pygmy-possum populations. Fortunately, many possums have been able to survive the fires by hiding in boulder fields. When they emerge from hiding, however, they encounter a new problem: the plants that supplied them with much of their food are gone, leaving them at risk of starvation.

All told, the mountain pygmy-possums are facing so many challenges that they are at high risk of extinction. If people want to intervene to save the mountain pygmy-possum, they must work quickly. As Archer puts it, "We don't have time to muck about here."

RESCUING THE
MOUNTAIN PYGMY-POSSUM

Australians know what happens if they fail to act quickly to prevent animal extinctions. "Unfortunately, in Australia, we do have the worst mammal extinction record of the world," Dr. Marissa Parrott, a reproductive biologist at Zoos Victoria told me. Her organization is working to ensure that "no Victorian bird, mammal, frog or reptile will ever go extinct on our watch." She explained that "in the past, people have monitored species into extinction, and you can see it with animals like the Bramble Cay melomys."

The Bramble Cay melomys were small rodents that became world famous in 2019 for being arguably the first known species to be declared extinct as a result of human-caused climate change—in this case, the extinction resulted in part from rising sea levels. The melomys were once inhabitants of Bramble Cay, a low-lying island in the northern part of Australia's Great Barrier Reef. When it rains, the island accumulates a pool of fresh groundwater that is needed to sustain the island's plants. During high tides and storms, saltwater sometimes pours into this freshwater reserve, which has resulted in the death of the plants that the melomys depended on for sustenance. Such saltwater intrusion events have become more common in the last few decades as the sea level has continued rising; by 2014, 94 percent of the island's vegetation had been killed by seawater. Without any source of food, the melomys starved into extinction.

The extinction of melomys was not a surprise, however; in fact, it was predicted well in advance. The population had been declining for years, and the problems caused by seawater inundation and its affect on the island's vegetation had been documented. Recognizing the risk of melomys extinction, the Australian government developed a plan to rescue the animals in 2008. Unfortunately, they took too long to execute their plan, and by the time they went to the island to collect the last remaining melomys for captive breeding in 2015, they could not find any; the species had already become extinct.

Advocates for the mountain pygmy-possum are trying to prevent the marsupial from suffering a similar fate by moving forward on three fronts at once. First, they're trying to help the possums adjust to changes occurring within their alpine environments. Second, they are breeding the possums in captivity, creating an option to house them in a perpetual ark. Finally, some advocates are proposing that wild mountain pygmy-possums be moved to new places, where they might have a better chance of surviving.

Getting used to it: phenotypic rescue

As with the tigers in India's Panna National Park, one strategy for saving species such as mountain pygmy-possums is to reduce local sources of death, including predation from introduced foxes and cats and collisions with cars on alpine roads. If this strategy is successful, it will increase the chances that more possums would be able to reproduce, preventing populations from disappearing in the future. And, indeed, this kind of action is already underway.

The Australian government released a species recovery plan for the mountain pygmy-possum in 2016 that describes a wide range of interventions, including removing foxes and feral cats, creating artificial rock piles to provide new habitat, fencing possum habitat to keep people out, and siting road development projects to avoid possum habitat. Tunnels have been built beneath roads on Mount Hotham and Mount Little Higginbotham to enable pygmy-possums to travel without becoming roadkill. Given the tunnels' role in facilitating mountain pygmy-possum trysts, they have been appropriately named "tunnels of love."

In the wake of increasingly intensive wildfires that destroy plants that provide much of the mountain pygmy-possums' food, wildlife professionals are also feeding the animals to keep them from starving. For example, Broome told me that during Australia's intense 2019–2020 bushfire season, a catastrophic wildfire burned a third of the pygmy-possum habitat in Kosciuszko National Park, removing all vegetation from the possum's boulder field habitats. To make matters worse, a drought in the lowlands had recently reduced the abundance of Bogong moths. Conditions were so extreme in the park's burned areas that Broome, a group of volunteers, and colleagues at the Department of Planning, Industry and Environment set up and maintained dozens of feeding and water stations to sustain the animals. They provided "Bogong bikkies," tiny energy bars for possums developed by Zoos Victoria and formulated to be nutritionally similar to Bogong moths. Broome shared the recipe with me: "mealworms, macadamia nuts, coconut and sunflower oil, and a whole lot of minerals and vitamins." The possums responded well to the feeding program, ultimately faring better than other possums in the unburned southern areas of Kosciuszko National Park.

While reducing mortality and feeding possums can temporarily stop or slow their rate of decline, whether or not the mountain pygmy-possums will ultimately persist in the mountains will likely depend on whether they can adjust their phenotype—their observable traits, such as appearance, physiology, and behavior—quickly enough to keep pace with environmental change. An organism's phenotype will change over its lifetime as it ages or encounters new environmental conditions. (For example, I experienced a change in my own phenotype recently when I went hiking in India's Himalaya Mountains at an elevation of about 13,000 feet [4000 m] above sea level. This altitude was much higher than I had ever been, and I had never experienced air this thin. My body, detecting the change, automatically turned on special genes to help me adjust by producing more red blood cells and hemoglobin in an effort to pull more oxygen out of the air. My body also signaled to me to slow down when my blood oxygen started to drop: my breathing became labored, my legs got tired, and my fingers started to tingle. So I sat down frequently to take breaks, breathe deeply, and let my oxygen levels recover. Aspects of my phenotype—in this case the composition of my blood and my behavior—were automatically shifting to help me survive stressful conditions at high elevation.)

All organisms have an innate ability to shift their phenotype when faced with different environmental conditions. I refer to this as *phenotypic rescue*, when an organism adjusts its phenotype to stay alive under conditions that would otherwise threaten its survival. If mountain pygmy-possums are to persist in their current populations, they'll have to adjust their phenotypes in many ways at once. For instance, they'll have to adjust to the new snow regimes, perhaps by storing more food for the winter or digging deeper burrows to escape the coldest temperatures in poor snow years. They'll have to adjust to fluctuations in the availability of Bogong moths and other food, perhaps by finding new foods to eat when their preferred foods are scarce. They'll also need to adjust to predation pressures from feral cats and foxes, perhaps by developing more effective predator avoidance behaviors. These kinds of phenotypic changes are very common in nature as species encounter new environmental conditions. The question is whether these pygmy-possums can adjust their phenotypes in many different ways over a very short period of time—and, so far, there's little evidence that they can.

Indeed, many mountain pygmy-possum experts are concerned that phenotypic rescue in the possums' alpine habitats won't be sufficient to save the animals. For example, in a 2019 article, Archer concluded the following: "It appears that under projected climate change scenarios extinction of this high-elevation specialist species in the wild is inevitable." Parrott thinks that the pygmy-possums can persist in the mountains only if they continue to be helped by people: "I think if we all just walked away and did nothing for the mountain pygmy-possums, they would go extinct." Given the uncertainty around the possums' future in the alpine, some advocates are already working on two backup plans: housing them in captivity and moving them to new places.

Living in an ark

Biologists at Zoos Victoria are currently raising mountain pygmy-possums in captivity. Captive breeding can act like an insurance policy for the species, so that even if the animals become extinct in the wild, they won't disappear from the Earth. Instead, mountain pygmy-possums could be relegated to existing only as captive curiosities of evolution and symbols of the biological consequences of climate change. And they wouldn't be alone. As more and more species are nearing extinction, including frogs, birds, reptiles, mammals, plants, and fish, some conservationists are choosing to save them in captive-bred arks.

Captive breeding of mountain pygmy-possums is also helping to maintain wild populations. For example, in 2013, when pygmy-possum populations on Mount Buller reached extreme lows, Zoos Victoria and its partners supplemented this wild population with thirteen captive-bred individuals for both demographic rescue (boosting overall numbers with immigrants from elsewhere) and genetic rescue (reducing the genetic problems of inbreeding). Maintaining species in arks also retains future conservation options, such as reintroducing them to the wild in the future if environmental conditions shift to more favorable states (thought it's not clear how that would be the case for mountain pygmy-possums). Captive breeding is an expensive undertaking, however, and this strategy will be limited to the few species deemed valuable enough to provide perpetual care. Furthermore, captive breeding alone is not a long-term solution to the problems that mountain pygmy-possums are facing in their mountain homes. Although the program may help prevent extinction and bolster wild populations in the near term, something more will likely be needed if mountain pygmy-possums are to persist in the wild.

Finding a new home: geographic rescue

Perhaps the most realistic option for saving mountain pygmy-possums is geographic rescue, in which animals relocate to avoid their extinction.

In theory, geographic rescue is one of the easiest ways for a species to deal with environmental change. For geographic rescue to succeed in nature, organisms require two key factors: a new location with a suitable environment and a way to travel to the new location.

Many species can move to new locations on their own, especially if they are good at dispersing. For example, alpine species all over the world are starting the process of geographic rescue by marching uphill to follow cooler weather as the climate warms. Unfortunately for mountain pygmy-possums, they have already found their way to the tops of their mountain homes, so moving to higher elevations isn't an option.

Alternatively, mountain pygmy-possums might be able to move to new mountains with more suitable alpine habitat. Connected protected area networks, like those that are helping to save tiger populations in India, can help promote this kind of species relocation as the environment changes, provided the species is sufficiently mobile. For many alpine species, however, movement to a new alpine area won't be easy unless they can fly or somehow hitch a ride from one mountain to the next. Earthbound alpine species such as mountain pygmy-possums would have to climb down their mountain, cross a valley, and then climb a new mountain somewhere else. Pygmy-possums are unlikely to make such a venture into unfamiliar and unsuitable habitat—and even if they did, they'd be at high risk from predators.

Could people intervene by moving the possums to new mountain habitats in Australia, sparing them these travel risks? Unfortunately, the answer is no, because no other alpine regions in Australia would provide suitable habitat for mountain pygmy-possums. To find sustainable new alpine habitat, people would have to move the possums beyond the continent's shores. For example, they could theoretically be moved to New Zealand, which has higher mountains than Australia and abundant alpine habitat. Such a move, however, would be considered radical and risky in conservation biology, in part because New Zealand has many unique and rare species of its own that might be harmed by introducing a new species.

In fact, New Zealanders have already experienced the negative effects of introducing another of Australia's possums. In the mid-1800s, European colonists deliberately introduced a distant cousin of the mountain pygmy-possum, the common brushtail possum, to New Zealand as a source of food and fur. Over the years, brushtail populations exploded, and today the animals have spread throughout much of New Zealand. The hungry brushtails are infamous for munching all the leaves off trees, in some cases killing them, and for competing with and consuming native birds. Broome noted that "mountain pygmy-possums might survive in New Zealand," but the New Zealanders "might be aghast at the thought, having experienced our brushtail possums over there."

With no domestic alpine habitat options available and considering the risks involved in moving mountain pygmy-possums abroad, a group of Australian conservation biologists has proposed an out-of-the box alternative: reestablishing the pygmy-possums in the lowland forests of southeastern Australia, where they once lived for millions of years. What would such a new life in lowland forests look like for mountain pygmy-possums? For this to work, the animals would need to undergo phenotypic rescue here as well, because they'd need to learn to eat different foods, avoid new predators, and adapt to a completely different climate. Because so many phenotypic adjustments would be required, it is uncertain that the pygmy-possums would survive such a move. However, researchers and conservationists led by Archer believe that the animals may be up to the task. When I spoke with Archer, he even went so far as to say, "I'm quite happy to put the left portion of any part of my body on the table as a bet that they will absolutely be happy little pygmy-possums" in the lowland forests.

Archer's confidence is based on the observation that mountain pygmy-possums evolved from lowland forest dwellers. In fact, most of the twenty-six-million-year fossil record of similar pygmy-possums indicates that the mountain-dwelling animals' ancestors lived in cool, lowland rainforests. Archer and his colleagues argue that today's alpine possums are still likely to have many of the traits that helped their ancestors succeed for so long at lower elevations, and they are trying to lessen the stress of the phenotypic adjustment by gradually introducing possums to lowland habitats.

When I spoke with Archer in late 2020, he and his colleagues were already constructing a lowland breeding facility, where pygmy-possums could live while they adjusted to a new environment. Within the facility, the animals will be introduced to new foods and other species that inhabit the lowland forests, in hopes of helping them adjust to the new habitat.

The possums will later be placed in a fenced outdoor enclosure designed to keep predators, such as foxes and cats, out. If they survive in the outdoor enclosure, the final stage would be releasing some of them into unfenced lowland wet forest habitats. To give them the greatest chances of survival, researchers would choose locations that offered sheltering rock piles that the possums prefer. They would also radio-tag each possum so that it could be tracked over time. If the effort is successful, the lowland-dwelling mountain pygmy-possums would breed, and their populations would grow and thrive in perpetuity.

As the mountain pygmy-possums' alpine environment continues to change rapidly, moving them to lowland forests may be the best option to avoid their extinction in the wild. The proposed relocation comes with risks, however. Current wild populations would be the source of possums for the breeding facility, and their capture would necessarily lower the abundance of already beleaguered wild populations.

Archer said that he and his colleagues have permits to collect twenty wild mountain pygmy-possums, which, at about 1 percent of the current wild population, is unlikely to cause much harm to overall numbers. The second risk is to the captive-bred animals, which may suffer in captivity and after release (especially if they have difficulty adjusting to life at lower altitudes).

Finally, the relocation could be so successful that it leads to unintended consequences, such as disruptions caused by competition with other species in the new habitat. Relocation advocates argue that this risk is low, however, because mountain pygmy-possums evolved in these forests and lived there as recently as the Pleistocene alongside most of the species that still inhabit the forests today. If the pygmy-possum's ancestors inhabited these forests recently (in a geological sense), how much harm could their reintroduction cause? The native species, especially ecologically similar small mammals, should be able to coexist successfully with mountain pygmy-possums, just as they did in the past. The only way to know for sure is to take a chance by purposefully relocating them.

Today, examples of human-assisted geographic rescue are rare in conservation, because of a general reluctance to take on the risk of something going wrong. But thousands of species have been introduced to new locations for non-conservation reasons—for agriculture, for aesthetics, and by accident. Sometimes introduced species have caused significant harm to, and even extinction of, native species, especially in confined spaces such as islands and lakes, where native species had no place to escape from the newcomers. In such cases, extinctions are even more likely if the native species are evolutionarily naïve to the new species, meaning that they have no recent evolutionary history of dealing with a new predator, competitor, or disease, which makes them particularly vulnerable. However, "foreign species hardly ever cause native species to become extinct from entire continents," according to ecologist and evolutionary biologist Chris D. Thomas from the University of York. Instead, if introduced species persist, they typically assimilate into the local ecosystem alongside the native species, increasing the local species diversity.

Within the field of conservation, discussions of whether or not to intervene in cases like the mountain pygmy-possum are exposing an internal conflict. One side suggests that moving a species to a new location is a risky strategy because of the unknown effects that the new species may have on the ecosystem, and they point to examples such as the introduction of brushtail possums to New Zealand or cats and foxes to Australia. The other side posits that not moving species to new locations is a risky strategy because it increases the risk of the species' extinction, and they point to species such as the Bramble Cay melomys. Both sides have valid points; the question is how can people weigh the opposing risks of relocating versus not relocating species? The mountain pygmy-possum is forcing conservation biologists to confront this internal conflict head-on with the future of the species hanging in the balance.

THE FUTURE OF MOUNTAIN PYGMY-POSSUMS

On its own, the rescue effect is probably insufficient for saving the mountain pygmy-possums from extinction, mainly because the nature of the possums' preferred habitat makes them inherently vulnerable. By being isolated in small populations on mountain tops, the animals aren't able to be rescued in the same way that tigers and salmon, which are capable of moving to new locations (geographic rescue) or exchanging individuals and genetic diversity between populations (demographic and genetic rescue), can. Instead, without human intervention, the possums must rely primarily on a combination of reproductive rescue and phenotypic rescue. Though both of these processes are certainly underway, there are good reasons to believe that without human assistance, the mountain pygmy-possum will face extinction, overwhelmed by the rate and magnitude of environmental changes.

Even with human intervention to boost the rescue effect, none of the conservation options for mountain pygmy-possums is ideal. Actively protecting them in their alpine homes risks failure if, as predicted, phenotypic rescue ultimately won't succeed. And this effort would be increasingly expensive, especially if possums continue to require deliveries of food and water along with other interventions, such as the culling of cats and foxes. Relegating them to captive breeding is also expensive and arguably less inspiring than having the species persist in the wild. Reintroducing them elsewhere would both test the limits of phenotypic rescue and raise a host of thorny questions about the ethics of human-assisted geographic rescue.

Most conservationists would argue that people have a moral obligation to save the mountain pygmy-possum because we are clearly responsible for its current plight. That said, whether and how to intervene on behalf of the mountain pygmy-possum will be informed by society's assessment of their value. Mountain pygmy-possums have no obvious commercial value, and they don't appear to have any major effects on their ecosystem. Determining whether efforts to slow their extinction are worth the cost is largely a question of the difficult-to-quantify intrinsic value of their existence. Broome noted that "the mountain pygmy-possum is pretty special in Australia, because it's the only marsupial that hibernates under the snow throughout the winter." Thus, by the very act of existing, the mountain pygmy-possum adds to the overall variety of our planet and our shared cultural inheritance of Earth's biodiversity. But this kind of argument can be made for any species—by definition, each species on the planet is unique in some way. In conservation, this argument goes only one way, whereby uniqueness merits conservation action. In all my years working in ecology and conservation, I've never once heard a conservation advocate say, "This species is not very special. We can just let it go extinct."

When I asked Broome about the reasons for saving the pygmy-possum, she included an extremely important one: "It's adorable, of course." Being adorable has helped the tiny possum develop a supportive constituency of wildlife managers, biologists, conservation professionals, and the public. So far, many conservationists have gone to fairly extreme lengths to save the species, including culling cats and foxes, modifying development plans, building tunnels, instigating captive breeding, and even serving Bogong bikkies.

Perhaps Archer and his colleagues are correct, and the best option for helping the mountain pygmy-possums survive is to relocate them to the lowland forests where their ancestors lived. For now, plans are moving forward to move some possums; the permits to collect the animals are in hand and the breeding facility will soon be ready for the possums to move in. If everything goes well, wild mountain pygmy-possums may once again occupy lowland forests in just a few years. But before that can happen, the New South Wales Parks and Wildlife Service will have to approve their plan, presumably by weighing whether the potential benefits of relocating the pygmy-possums outweigh the risks.

Like it or not, the mountain pygmy-possum is a harbinger of what's to come in conservation. As climate change intensifies, more species will end up trapped on their metaphorical mountain tops, in fragments of habitat that are growing increasingly unsuitable and isolated from other places that might better suit them. The relocation of mountain pygmy-possums is an early test case to determine how conservation should advance in a changing world. Should people accept the loss, or indefinite captivity, of a species that cannot survive on its own? Or should we step in, accept the risks of assisting in geographic rescue, and play god by deciding which species will be relocated in hopes of saving them?

CHAPTER 4

A Babe
in the Woods

IT IS NEW YEAR'S DAY, 2021, and the world is in the grips of the COVID-19 epidemic, but I'm thinking about a different epidemic that started more than a century earlier. On my computer screen is a picture of William Powell, a professor and director of the American Chestnut Research and Restoration Program at the State University of New York's College of Environmental Science and Forestry. He has graying hair and glasses, and he's wearing a comfy jacket, jeans, and hiking shoes. He's standing in a field with rows of small trees. A shovel is stuck in the ground to his right, and in front of him is a perfect hole ready to be filled by a potted American chestnut tree to his left.

The little tree is not much to look at, with ten or so droopy, kayak-shaped, pointed leaves. It is tied to a stake to keep it upright in the pot, though it hardly looks to be in any danger of blowing over. Although it's not clear from the photo, this unassuming tree is actually a miracle of human bioengineering, because it can do something that American chestnut trees have struggled to do for more than 100 years: it can resist the deadly chestnut blight disease.

Powell's tree is the result of a new and—depending on your point of view—extraordinarily elegant or frighteningly unnatural approach to wielding the rescue effect to save a species. It was genetically engineered to include a new gene from an entirely different plant, which helps it resist chestnut blight. That makes it one of the very first genetically modified organisms (GMOs) to be created in the name of conservation. The tree's very existence is controversial to people who think genetic engineering crosses an unacceptable ethical line, and yet it may hold the key to rescuing not only American chestnut trees, but many other species that will need to evolve to persist in a changing world.

DEAD AND DYING TREES

In 1906, Hermann Merkel, chief forester of the New York Zoological Park (now the Bronx Zoo) was alarmed. Merkel was responsible for looking after the park's vegetation, and the American chestnut trees in the park were suffering from an unknown disease. He wrote the following:

> This disease was first noticed in the New York Zoological Park, in a few scattered cases which occurred during the summer of 1904. Since that time, however, it has spread to such an extent that to-day it is no exaggeration to say that 98 per cent of all the chestnut trees in the parks of this Borough are infected. The spread of this disease is so sudden that unless some radical

measures are taken, or a natural enemy of this fungus develops, it is safe to predict that not a live specimen of the American Chestnut (*Castanea dentata*) will be found two years hence in the neighborhood of the Zoological Park.

At the time, the American chestnut was one of the most valuable and abundant trees in the forests of the eastern United States. In an attempt to save the park's trees and stop the spread of the disease, Merkel removed infected branches and burned them, cut down heavily infected trees, and even tried dousing trees in antifungal chemicals to kill the blight. None of these efforts was successful in containing the disease, and the chestnut blight began spreading like wildfire across New York and nearby states.

The blight reached Connecticut, Massachusetts, and Washington, D.C., by 1908. William A. Murrill, one of the first scientists to study the disease, concluded, "If this disease continues as it has begun, there is, theoretically, no reason apparent why it should not sweep from the country practically every tree, both native and cultivated, of the genus *Castanea*. Let us hope, however, that, in the economy of nature, something will intervene to prevent this."

By 1912, the chestnut blight was present in at least ten states, and some were taking active measures to stop it. Among the most aggressive actions were those undertaken in Pennsylvania, which established the Chestnut Tree Blight Commission to find and cut down any diseased trees. By the next year, however, it was already clear that these efforts were woefully ineffective. Commission Chairman Winthrop Sargent concluded that, "The complete loss of the present commercial stand of chestnut in Pennsylvania . . . seems absolutely certain."

Sargent was right—and not just about Pennsylvania. Over the next few decades, the blight progressively spread throughout the range of the American chestnut, indiscriminately killing trees. By the 1920s, the blight was marching southward along the Appalachian Mountains. As the blight approached North Carolina in 1923, plant pathologist G. Flippo Gravatt lamented, "When from a mountain top one looks over thousands of acres of vigorous chestnut, it is indeed hard to believe that within a few years the view will be changed to one of dead and dying trees."

By 1940, the spread of the disease was nearly complete and an estimated three to four billion chestnut trees were dead. "We cannot verify that any stem survived the original pandemic," said Sara Fern Fitzsimmons, director of restoration for The American Chestnut Foundation. In other words, as far as we know, not a single American chestnut tree was left standing.

THE PERFECT TREE

My first encounter with the American chestnut was through a relative's Connecticut home renovation project. I remember his excitement as he told me about how he had discovered chestnut planks under the home's old floor during the demolition. His house had been built in the early 1800s—long before the blight—and chestnut would have been a relatively cheap wood for construction at the time. He considered this find to be particularly valuable specifically because it was American chestnut, even though the wood looked rather knotty to me. He had the planks painstakingly removed, refinished, and reinstalled in the remodeled kitchen. I noted his enthusiasm and chalked it up to the fact that he was a professor in a university forestry department and an avid woodworker. At the time, I didn't know the story of the American chestnut, so I incorrectly assumed that his views were those of a specialist. I now know that his views were part of a more broadly held yearning for the return of this lost tree.

The American chestnut wasn't just any tree—it was deeply embedded within American culture and was extremely valuable economically. Once widespread, the trees ranged from Ontario, Canada, to Mississippi and west to the Ohio Valley. Throughout much of that range, the trees were often one of the most abundant trees in the forest.

In *American Chestnut: The Life, Death, and Rebirth of a Perfect Tree*, author Susan Freinkel described the "perfect tree," with "a utilitarian versatility no other tree could match." Though most American chestnuts were growing in native forests, the tall and stately trees were also planted in formal gardens. The fast-growing hardwood featured a straight trunk and was valuable for making furniture, power poles, railroad ties, fences, and, apparently, kitchen floors. Freinkel noted that, "At the peak of its production, between 1907 and 1910, chestnut wood contributed more than ten million dollars annually to the economy of Appalachia"—the equivalent of about $282 million in 2020 dollars.

The trees also provided a reliable and salable crop of nutritious tree nuts that Henry David Thoreau called "a good substitute for bread." Anyone could collect the ripe nuts off the ground in the fall. For many rural families, they provided a rare opportunity as a cash crop to earn money, and even city dwellers would make excursions to the countryside to go "nutting." Native Americans saw similar food value in chestnuts, in part because, unlike acorns, they could be eaten without having to be boiled for hours. Some Native Americans even burned down competing tree species to make chestnut orchards.

In addition to offering many benefits to people, the American chestnut was an integral part of forest ecosystems. Many species of wildlife relied on the nuts as a source of food, as deer, bears, passenger pigeons, and wild turkeys descended upon chestnut groves for an annual feast.

Several species of moth relied exclusively on the American chestnut tree to complete their lifecycles.

Though very few people living today have witnessed the great American chestnut forests of the early twentieth century, a vibrant constituency remains dedicated to restoring them to their past glory, and in so doing, reclaiming some of the values they once provided. Lisa Thomson is president and CEO of The American Chestnut Foundation, a nonprofit with sixteen chapters across the country, whose mission is "to return the iconic American chestnut to its native range." She told me that many organization volunteers are retirees who offer their time and resources to efforts to restore the tree. They are "obsessed in a good way," she said, adding, "It's a true labor of love."

THE RISE AND FALL OF
A BABE IN THE WOODS

As the chestnut blight was infecting and killing billions of trees, people naturally began to wonder how the disease appeared in New York in 1904, and why was it so devastating to the American chestnut? The answers to these questions lie in the evolutionary history of trees in the Northern Hemisphere and the consequences of highly mobile people who have a knack for rearranging the diversity of life on Earth in ways that nature never envisioned.

Eurasia and North America are separated by two vast oceans, yet they share many similar species of trees. For example, most people from the eastern United States, in addition to those from France or China, are familiar with oaks, maples, and beeches, because these trees grow in those places. How did this similarity in tree types arise across these disparate locations? Many scientists believe that, in the Mesozoic Era during the time of the dinosaurs, a great forest extended across the northern reaches of today's North America, Asia, and Europe. Within this continuous forest were the ancestors of modern species of northern trees such as maples, oaks, and chestnuts. When the climate changed, the trees migrated south (through geographic rescue), where they became isolated from one another by oceans and other barriers such as mountain ranges.

The ancestors of modern chestnut trees were eventually divided into separate populations in today's Europe, Eastern Asia, and eastern North America. Once separated, the trees experienced differences in climate, soils, competitors, and pests such as herbivores and parasites. Under these different conditions, they started to evolve slowly into different species, as survival of the fittest favored individual trees with genetic combinations that thrived in each location. Over time, the trees in different locations developed different leaf shapes, nut sizes, and growth forms.

In evolutionary biology, this process is called *allopatric speciation*, and it occurs when new species arise after populations are separated into different locations. Allopatric speciation is believed to be the primary generator of the diversity of life on Earth. Whenever a population is divided, each separate population has a chance of evolving into its own unique species. Then one ancestor species gives rise to two sister species, which are similar to each other, but not quite the same.

The process of allopatric speciation also occurred within continents when tree populations were separated for long periods of time. Scientists are still studying chestnuts to determine exactly how many species were ultimately produced though this process. At least two chestnut species are native to North America: the American chestnut (*Castanea dentata*) and the Allegheny chinkapin (*C. pumila*). Europe has one native species, the European chestnut (*C. sativa*). Four native species are found in Eastern Asia: the Chinese chestnut (*C. mollissima*), Japanese chestnut (*C. crenata*), Seguin's chestnut (*C. seguinii*), and Chinese chinkapin (*C. henryi*).

As the various species of chestnut were evolving separately in different locations, some of the populations would have encountered new pathogens. In response to a new pathogen, an infected host will usually mount a defense through its immune system with the aim of lessening or eliminating harm. This change in physiology is a form of phenotypic rescue, and this is what has happened with the COVID-19 virus and humans. Most infected people haven't died from the virus; instead, through phenotypic rescue, some immune systems can eventually overpower the virus and the person survives. Human immune systems can even store information about the virus within white blood cell antibodies, so that their immune systems can recognize and attack the virus more quickly in the future. As a result, people who have survived COVID-19 may develop some immunity to future infections. This is also how vaccines work: by tricking the immune system into developing the right antibodies, an immune system knows to attack the virus immediately if the host is infected.

At some point in the distant past, Asian chestnut species encountered a fungal pathogen, *Cryphonectria parasitica*, which causes chestnut blight. The fungus is a wound parasite that can attack trees only through a cut or tear in the bark, which might be caused by a scrape from an animal claw or a broken branch. It then spreads into the inner tissues of the tree.

Ancient Asian chestnut species must have initially mounted a successful phenotypic response after they were infected, which enabled them to survive the first waves of infection. Over many years, through eons of natural selection, the Asian species accumulated specific genes that helped it fight chestnut blight. As the trees evolved, the fungus was probably simultaneously evolving newer and better ways to get past the trees' defenses. The net result was a co-evolutionary arms race between chestnut trees and the fungus, each side incrementally evolving to deal with the other more successfully. Through evolution, Asian chestnut trees and C. *parasitica* have arrived at a détente, whereby the Asian trees can usually contain the damage caused by the fungus but cannot defeat it entirely.

In the years prior to the emergence of chestnut blight in New York, people in the United States had been importing exotic chestnut trees from Asia for their gardens, yards, and farms. Apparently, some of these imports were carrying an unexpected stowaway: C. *parasitica*. As the fungus grew on the imported trees, it would have eventually turned its attention to reproduction.

Many fungi have mushrooms, which are used to disperse reproductive spores. The C. *parasitica* fungus doesn't produce anything that looks like a typical mushroom, however; instead, it produces small, orange to yellow bumps that erupt through the tree's bark and release spores. Some of these spores are excellent at dispersal, with the ability to be carried by the wind or to hitch a ride on the bodies of birds, insects, or mammals. By chance, some of the spores produced on imported Asian chestnut trees must have found their way into wounds on some American chestnut trees, which launched the epidemic.

Fungi are neither plant nor animal and belong to their own separate biological kingdom, which also includes molds, mildews, and yeasts. Most fungi are decomposers, and their primary food source comprises both living and dead organic matter. The bulk of a tree's trunk is composed of dead wood in the middle of the trunk, and bark, which is also dead, on the outside. Sandwiched between the two are a few living layers that control functions such as circulating food and growing new wood. These crucial living layers are attacked, killed, and digested by C. *parasitica*.

When an American chestnut tree is first infected, the only signs are slight discolorations on the bark, above where the fungus is growing. As the fungus continues to grow, a canker develops on the bark. Underneath the canker, the fungus is killing and digesting tree tissues, creating a hollow space into which the bark sinks. At this stage, disease-resistant species, such as Chinese and Japanese chestnuts, can control the fungus by building physical barriers to prevent its spread, while simultaneously attacking it chemically. However, the American chestnut, a mere babe in the woods, has essentially no natural ability to resist the super-fungus that had spent untold generations fine-tuning its ability to attack and consume Asian species of chestnut trees.

As the fungus continues to spread under the bark, it often makes its way around a living branch, killing a complete ring of tissue. Because the infected branch is effectively cut off from the rest of the tree, it can no longer receive nutrients, so it withers and dies. When the infection attacks the trunk, it affects the entire tree, which can quickly die.

Shortly after the first American chestnut trees were infected, the fungus would have produced billions of spores, which then spread to nearby trees. As spores made their way into wounds on more and more chestnut trees, the spread of the blight accelerated. Because the trees have no natural disease resistance, the spreading wave of fungus-induced tree death that continued for decades was inevitable.

RESCUING THE AMERICAN CHESTNUT

To be successfully rescued, American chestnut trees need to be able to defend themselves against the fungus, or they need to avoid getting infected in the first place. Phenotypic rescue, via an immune system response, is often the best option for rescuing a species that encounters a new disease. However, because the immune response of the American chestnut wasn't able to defend against the blight, it would need to be rescued in some other way.

When a disease devastates a population of living organisms, reproductive rescue may be the best solution for restoring the population. When the chestnut blight killed most of the American chestnut trees, the fungus could have simply run out of new hosts to infect, causing it to disappear on its own. If that had happened, a massive boom of young American chestnuts may have emerged from the chestnut seedbank that contained billions of chestnut seeds spread across millions of acres of forest floor. Once sprouted, these young trees may have had unusually high survival rates, because many would be growing in unshaded places where the adult chestnut trees had recently died.

But a chestnut boom never happened, because the chestnut blight spread to new trees such as oaks. Unlike the American chestnuts, the oaks of the eastern United States were naturally resistant to chestnut blight, so they quickly formed a phenotypic détente of their own with the fungus. As a result, even after all the adult American chestnuts were gone, the fungus had become a permanent fixture in eastern forests, ready to infect any new chestnut tree sprouts. When a chestnut seed germinated and started to grow, it was just a matter of time until it, too, became infected, literally killing any chance that the chestnut seedbank could rescue the species.

Reproductive rescue even got a second—and somewhat unusual— shot at bringing back the American chestnut: many of the trees that appeared to have been killed by the blight were still clinging to life. Their branches, leaves, and trunk were indeed dead and gone, but their underground root systems, which the blight does not attack, remained alive. To this day, many of these root systems continue trying to grow new chestnut trees by sending up new shoots. Sara Fern Fitzsimmons told me that "we're now sitting at about 430 million" of these resprouting root systems across the eastern United States. However, as with the newly germinated trees, each time a sapling emerges, it is invariably infected with the fungus before reaching sexual maturity. These infected sprouts then become new sources for fungal spores, reinfecting any future sprouts. According to Fitzsimmons, the result is a repeating cycle, whereby new trees sprout, "they die back, they resprout, they die back, they sprout." On the rare occasion that a tree shoot beats the odds and grows to maturity, it still can't set seed, because a nearby tree is required to complete pollination, and the chances of two nearby trees reaching reproductive maturity at the same time are consistently low. For the past century, even with hundreds of millions of living root systems, the American chestnut has been unable to make a comeback.

Geographic rescue, however, may be a viable option for the American chestnut. The trees cannot travel to new locations on their own, of course, but people have planted the trees in new locales, such as Oregon and Wisconsin, with some success. Their survival there may be temporary, however, as the fungus has already spread to the trees in Wisconsin. All is not lost, because some locations may be unsuitable for the fungus to survive, and this could offer the trees a long-term refuge. American chestnuts could be geographically rescued with the help of conservationists who plant them in locations where the trees can survive but the fungus cannot. But even if this is possible, geographic rescue does not create an attractive solution for many advocates of the American chestnut, who want to restore the tree as a dominant species in the forests of the eastern United States.

Another possible solution might be to replace American chestnuts in US forests with a blight-resistant species of chestnut to provide benefits such as nuts and lumber. The best candidate for this might be Asian chestnuts, which are currently grown in yards and orchards. This is not a feasible solution, however, because the relatively short Asian chestnuts would not be able to compete with taller species such as hickory, maple, beech, and others in these forests.

Ultimately, restoring the American chestnut to its former range would likely require evolutionary rescue, which occurs when organisms successfully evolve their way to being able to survive under stressful conditions. Most people think of evolution as a slow process that can take millions of years. This long view is appropriate for major evolutionary changes that require a slow accumulation of millions of small changes that arise haphazardly in DNA, such as humans evolving from monkey-like ancestors. But evolutionary rescue can occur very quickly if two conditions are present: First, the organism in question must experience *differential mortality*, which refers to the fact that some organisms will die because of a particular threat such as a new disease, while others will survive. Second, the population must have enough *heritable genetic variance* with regard to the trait in question—in other words, the survivors will have genetic combinations that are better at dealing with the threat, and they can pass those genetic combinations to their offspring. If these two conditions are met, the genetic traits of one generation can be different from their forebears—this is evolution.

At the outset of the blight epidemic, the American chestnut may have appeared to be a decent candidate for evolutionary rescue. Over its evolutionary history, the species has adapted to survive in a variety of habitats, ranging from Canada to the American Deep South. In doing so, it developed a reservoir of genetic diversity, whereby populations in one area were genetically distinct from those in another area. As the fungus began moving through these populations, it might have encountered some trees with genetic combinations that enabled them to resist the disease. Because of the enormous numbers of American chestnuts growing in North America, there were billions of tickets to this evolutionary sweepstakes.

Unfortunately, the right genetic combinations did not exist. Of the billions of trees lost, Fitzsimmons told me, "We've only found a couple dozen trees that actually have some measurable amount of resistance to the blight. . . . Every other tree that we come across has zero resistance." Although the chestnut's ability to resist the blight occurs along a gradient, even the most resistant wild trees eventually become overwhelmed by the blight and die.

Ultimately, the chestnut blight systematically defeated all the processes that make up the rescue effect. Reproductive rescue was impossible because the trees died before reaching sexual maturity. Demographic rescue and genetic rescue couldn't help because the disease had already spread to all populations and killed every standing tree. Natural geographic rescue wasn't an option because it wouldn't restore the trees to native forests, and the blight would likely follow the trees to new locations. And neither phenotypic rescue nor evolutionary rescue could save the trees. Realistically, that leaves only one last option for the American chestnut: boosting its ability to resist the blight.

ASSISTED EVOLUTION

Within the field of conservation biology is a new branch that is vigorously growing, called *assisted evolution*, in which scientists aim to speed up the process of evolution for species that are struggling to keep up with environmental change. In other words, assisted evolution can help boost evolutionary rescue.

People have been manipulating the evolution of domestic species since the dawn of civilization. For example, through selective breeding, people have produced numerous varieties of plants and animals—including corn, wheat, potatoes, cattle, and dogs—that can survive new conditions in the service of human needs. As the chestnut blight was raging through the eastern United States, agricultural researchers began exploring various methods for breeding blight-resistant trees.

The most conceptually straightforward way to breed a blight-resistant American chestnut would be to breed blight-resistant survivors. Some groups, including the American Chestnut Cooperators' Foundation, are trying to do this through cross-pollination of trees that resprout. However, they are encountering the same problems experienced in natural evolutionary rescue: the genes needed to overcome the pathogen just don't seem to exist in wild American chestnut trees. And breeding has yet to produce trees with enough resistance to warrant large-scale replanting efforts.

Given that the requisite genes don't appear to occur naturally, breeders are searching for ways to find new genes from other species to introduce to American chestnuts. The most obvious places to look for these genes are in the Asian chestnut species, which are already resistant. Within decades of the first appearance of the blight, breeders began working on hybridizing American chestnut trees with other species, in hopes of finding the key to resistance. Hybrids are not members of either parents' species; they are new organisms that include genes from both. Perhaps the most famous example of a hybrid is the mule, which is the offspring of a female horse and a male donkey.

During the last century, breeders have hybridized thousands of surviving American chestnuts with Asian chestnut species, especially Chinese chestnuts. These efforts, which began in the 1920s, continued for decades and produced some hybrids with better resistance than wild

American chestnut trees. However, the hybrids never reached a desired trifecta of traits—fast growth, blight resistance, and a tall, straight trunk—which are valuable for forestry and necessary for the trees to compete with other trees in the forest.

Early efforts to breed blight-resistant trees were shots in the dark, because they largely relied on first-generation hybrids, which were genetically half American chestnut and half Chinese chestnut. This strategy was unsuccessful because the blight-resistance genes from the Chinese chestnut parent were always diluted by half. To be fully resistant, the hybrids probably needed to carry two copies—one from each parent—of many of the resistance genes. So how could breeders create a resistant hybrid that still possessed the desired qualities of American chestnut trees?

As the twentieth century wore on, scientists developed tools for cracking this nut systematically through backcross breeding. To demonstrate this process, consider an analogous—but hypothetical—example of trying to breed a striped lion. Lions don't naturally have distinctive striping, but tigers do, and these two species can be crossed to produce a hybrid animal, a liger, which is genetically half tiger and half lion. Ligers typically have stripes, but they don't tend to be as distinct as a tiger's stripes, probably because the tiger stripe genes are diluted by half in the liger.

Starting with striped ligers, a breeder could start a multigenerational breeding program that attempted to make the hybrids more lion-like overall, while retaining only the stripes from tigers. To do this, the breeder would select liger cubs with the strongest striping patterns, raise them to adulthood, and breed them with normal lions. This process is *backcrossing*: a hybrid with a desirable trait (stripes) is bred back to a pure line of one of the parent species (lions). Backcrossing doesn't always work because of genetic problems with the hybrids, but for the sake of this hypothetical, let's assume that it does work with these lions and tigers. With each successive generation of backcrossing, the offspring would be genetically more and more lion, and less and less tiger. The tiger genes would become more dilute with each backcross, so the chances of losing the genes for stripes would increase. The breeder would have to produce enormous numbers of cubs in each generation, with the hope of finding a few that had more distinct stripes than the rest.

Once the breeder had completed enough generations of backcrossing, the next stage of breeding would be crossing the backcrossed hybrids. Pure lions would no longer be used in the breeding; instead, only striped lion hybrids would be bred with each other. If the backcrossing were successful, both parents would now have some of the genes for stripes, so their offspring could inherit two copies of these genes. However, if the presence of stripes is controlled by many genes, the breeder would have to produce huge numbers of cubs and hope that a few would have just the right combination of tiger genes for stripes from both lion-like parents. Most of these offspring would have intermediate striping, but once in a blue moon, a cub might be born with the stripes of a tiger on the body of an animal that was mostly a lion. These rare individuals would then become the brood stock for future generations of lion-like cats with stripes.

Since the 1980s, breeders have been trying to backcross a blight-resistant chestnut tree in a similar way. However, this process is more difficult than that of our hypothetical striped lion, because trees don't visually display traits for disease resistance. So breeders first have to raise young trees to a level of maturity sufficient to test them for resistance. The trees are then injected with the blight, and the breeder waits to see what happens. One cycle of this backcrossing process can take up to seven years.

With each generation, breeders select the trees with the best resistance. The others are culled, what Lisa Thomson euphemistically described as being "voted off the island," because of their low resistance to the fungal infection. Breeders of The American Chestnut Foundation have been breeding hybrid trees through backcrossing since 1983, and the trees they've crossed are now in their seventh generation. "It's slow. I mean, they're trees, right?" explained Thomson. So far, they've had some success, but she conceded, "Although we used the best science at the time, and we have some promising trees with intermediate resistance, we didn't have quite the blight resistance that we had hoped by the time we got to this point in the breeding program."

The process is taking longer than anticipated because of the number and complexity of genes involved. When the foundation's breeders started the backcross breeding program, they thought the disease-resistance of Chinese chestnuts might be controlled by just two or three genes. Although it would be difficult to produce a backcrossed, disease-resistant hybrid that had these two or three controlling genes, they believed it was doable. But as researchers closely examined the resistance genes in the Chinese chestnut, they found more genes. They now know that there are at least twelve resistance genes, and these are probably interacting together in complicated ways. When I spoke with Fitzsimmons in 2021, she told me that getting to a fully resistant and diverse population through backcrossing with this number of genes is not feasible, because it would require breeding "an unfathomable number" of trees in every generation—"more than we've planted in thirty-five years of breeding." Although they can get trees with partial resistance through backcrossing, the hybrid trees may never be as good as the Asian species at resisting the blight. It appears that backcrossing hybrids alone won't bring the American chestnut back anytime soon.

GENETICALLY ENGINEERED
CHESTNUT TREES

In the late 1980s, Herbert F. Darling Jr. and Stanley Wirsig, founding members of the New York Chapter of The American Chestnut Foundation, began considering options beyond backcross breeding. At that time, scientists were expanding the range of tools available for breeding organisms to include directly manipulating their genetic code. In 1989, Darling and Wirsig contacted professors William Powell and Charles Maynard at the State University of New York College of Environmental Science and Forestry. As Powell recalls, they wanted to know whether there was a way they could "use some of this newer technology—genetic engineering—to make a blight resistant chestnut tree."

Multiple terms can be used to describe what Darling and Wirsig were proposing—genetically modified organism (GMO) is probably the one most familiar to nonscientists. When people think of a GMO, they imagine an organism whose DNA has been manipulated by scientists. Many scientists don't use this term, however, because it could theoretically refer to any organism whose genetic makeup has been modified by another organism, including through classical breeding. Instead, scientists often use the term *transgenic* to describe an organism whose DNA has received genetic material from another organism through the process of genetic engineering.

Critics sometimes equate genetically engineered organisms to Frankenstein's monster—for example, some refer to genetically modified Atlantic salmon as "Frankenfish." This evokes a dark laboratory with a sinister scientist who is crossing ethical lines to create something inherently unnatural and dangerous. As a result, many people have a negative reaction to the concept of a GMO. Critics may have a valid point in that genetic engineering crosses a bright line in plant or animal husbandry. Selective breeding has always been a game of chance. For example, a breeder might get lucky while crossing American and Chinese chestnuts and create something new that has just the right mix of desirable traits.

Genetic engineering is meaningfully different from classical breeding because it allows for precise manipulations of targeted genes within an organism's DNA. These new genes can be created in the lab, or they can be plucked from another organism. Genetic engineering, in and of itself, is not inherently unnatural, however, because laboratory genetic engineering is merely an example of biomimicry, whereby scientists copy a process that was already invented by nature.

For example, *Agrobacterium* bacteria use genetic engineering to trick plants into making their food. To visualize how this bacterium genetically engineers plants, imagine a plant's DNA as a three-ring binder filled with recipes, called genes, each of which lists instructions for an important biological function. Machinery in the plant's cells reads these recipes and follows the instructions to make biological products—usually proteins, which are the building blocks of life. *Agrobacterium* have naturally developed genetic engineering tools to open the three-ring binder of certain plants just enough to insert a few new recipes for making its food. Once this is done, the plant cells will always have copies of the *Agrobacterium* genes in their DNA and will keep making the bacteria's food. The plant is now transgenic, with the *Agrobacterium* as the genetic engineer. In 2015, scientists discovered that this phenomenon has actually occurred in sweet potatoes—the kind you buy at your local grocery store—which have been naturally genetically engineered by *Agrobacterium*.

These bacteria are not the only natural genetic engineers. For example, a retrovirus, such as the human immunodeficiency virus (HIV) that causes AIDS, inserts its DNA into a host, tricking it into producing more copies of the virus. As scientists look more closely at the DNA of more species, they're finding some that are unexpectedly transgenic, such as aphids that have genes from a fungus and ferns that have genes from hornwort plants.

The transgenic organisms that have been the subject of significant controversy are not the naturally produced ones, but those that are designed by scientists. One of the ways that people genetically engineer organisms is by usurping the recipe-insertion machinery of *Agrobacterium* bacteria. With modern genetic tools, scientists can remove the genes that the bacterium inserts into its plant host's DNA and replace them with new genes of their choosing. One of the first successful examples of this was described in 1986, when scientists inserted a gene from a firefly into a tobacco plant. This early example of genetic engineering served no economic or functional purpose, but it made for an unmistakable demonstration that the gene insertion worked: under the right conditions, the tobacco plant glowed in the dark like a firefly.

With rapid improvement in the technology to genetically engineer organisms, Powell and Maynard decided to explore whether they could insert new genes for blight resistance into American chestnut trees. Instead of the scattershot method of hybridizing two species and then backcrossing, genetic engineering would enable their team to insert exactly the genes they wanted. But which genes should they insert?

Powell was in charge of identifying which genes might confer resistance so that his team could try to insert those genes—and only those genes—into the American chestnut. The most obvious place to look was in the DNA of Asian chestnut species. In the early 1990s, when he began these efforts, the available tools to read an organism's DNA and decode the target genes were still developing. As a result, he didn't know which genes conferred resistance in Asian chestnut species—and he believed several genes were involved—so he ultimately chose a different, and potentially easier, route: he looked for a single gene from any organism that could fight fungal pathogens.

Powell found a promising candidate in frogs that produce small proteins in their skin that help them fight fungal infections, so he and his team studied how these proteins affected the chestnut blight. They were not quite satisfied with the natural frog proteins, so they created genes of their own design based on the frog's genetic code. Bingo! When their new frog-inspired genes were added to American chestnut DNA, the result offered some blight resistance. Although the resistance was not strong enough to save the trees from blight, it was a promising start.

Perhaps by tinkering with this gene, or adding another, Powell's team might have created a tree with sufficient blight resistance. But there was a problem—not a scientific problem, but a perception problem. As Powell and his colleagues shared their results with members of the public, they found that people didn't like the idea of an American chestnut with "frog genes." This genetically modified American chestnut was triggering a broader set of social concerns about the ethics of genetic engineering. As a result, the scientists decided to abandon the froggy solution and look elsewhere for a gene that would confer resistance and would be easier to digest by the public.

Powell later wrote that he had his "eureka moment . . . while reading a book of abstracts from the 1997 Annual Meeting of the American Society of Plant Physiologists." As the blight fungus grows underneath the tree's bark, it sends out a chemical advance party, whose key ingredient is oxalic acid, which kills the tree's living tissues to prepare them for digestion. If the tree could somehow control the oxalic acid or reduce its effects, it might be able to resist the fungus. Powell knew that many plants suffer from fungal diseases, and in response, many of them have evolved defenses against oxalic acid. For example, corn, spinach, strawberries, rice, bananas, wheat, and cacao, naturally make an oxalate oxidase enzyme that chemically breaks down oxalic acid, neutralizing it. Powell and his colleagues concluded that this gene and its product, which is found in many foods, would seem less unnatural to people who might be uncomfortable with transgenic organisms. So his team copied the wheat gene for oxalate oxidase and inserted it into an American chestnut tree's DNA. It worked.

Throughout his many years working with the American chestnut, Powell has tried inserting many different genes to help it fight the blight. Although nothing has worked as well as the oxalate oxidase gene from wheat, it's not a perfect solution. His transgenic American chestnuts are still infected by chestnut blight, but they are more likely to fight the infection to a draw, with both host and pathogen surviving. In addition, because of a trick of genetics, only half of the offspring of transgenic American chestnut trees inherit the oxalate oxidase gene, so the other half are highly susceptible to the blight. Still, the trees that inherit the gene appear to have stronger blight resistance than the backcrossed hybrids. They're also not a messy mélange of two hybridized chestnut species; instead, they have all 30,000-plus American chestnut genes, plus the single gene from wheat.

RESTORING THE AMERICAN CHESTNUT

Now that blight-tolerant transgenic trees have been created, it might seem possible to begin restoring American chestnut forests in earnest. However, the few transgenic trees created to date include only a tiny fraction of the genetic diversity that was present in the preblight chestnut forest. This pool of historic diversity served at least two important functions: It enabled populations of American chestnut trees to adapt to local conditions, which varied hugely across the tree's native range. And the high genetic diversity helped give the species more genetic options when dealing with new environmental challenges, such as climate change.

Accordingly, Powell and his colleagues, plus dozens of citizen scientists from The American Chestnut Foundation, are starting to increase the gene pool of blight-tolerant American chestnuts by crossing transgenic trees with wild survivors. The offspring of these crosses will include a mix of locally adapted genes, and half of them will inherit the gene for oxalate oxidase. Using this approach, locally adapted and genetically distinct lines of American chestnut trees can be developed to match the places where trees have managed to survive amid the blight. This process will be a sort of preemptive genetic rescue, whereby new populations are deliberately founded with a mix of genetic combinations, reducing the chance of inbreeding in early generations.

Planting the genetically engineered American chestnut trees may begin in earnest during the next decade. If that happens, the American chestnut may be poised to make a grand return after a century-long march to extinction in its native forests. However, this scenario could be problematic, because the forests that were once home to the American chestnut have changed during its absence. And perhaps, most significantly, the climate has changed, with new temperature and precipitation regimes.

"The southern part of the range is being removed from the chestnut... because it is getting too hot for them down there," Powell told me. This means that the planned crosses of blight-tolerant and local trees may be starting off on the wrong foot, climatologically, and evolutionary rescue may be most effective if it's combined with a form of human-assisted geographic rescue, whereby the offspring of more southerly trees are purposely moved northward to match their preferred climate. Although it's uncertain whether the American chestnut will ever be restored to the southernmost parts of its historical range, the existing genetic diversity of the remaining southern trees can be used to kick-start new populations with traits that may help them survive in today's, and tomorrow's, climate.

TRANSGENIC CONTROVERSY

Before moving forward with restoring forests with transgenic American chestnut trees, a key question needs to be addressed: Is it acceptable to propagate them? To some, the answer is no, because the genetically altered trees themselves may pose too many risks. In fact, the Food and Agricultural Organization of the United Nations has listed eleven separate potential risks of transgenic organisms in agriculture across three categories: economic, human health, and environmental.

For the American chestnut, the economic risks are probably low, because the trees are not being developed for the exclusive economic purpose of a private company. Instead, they are being bred to share freely with anyone pursuing forest restoration. In addition, commercial harvest of American chestnut wood and nuts could effectively re-create an economic opportunity that was previously lost.

But could the transgenic American chestnut cause environmental problems? By choosing a common plant gene for translocation, Powell was aiming to address these kinds of environmental and health concerns from the start. This gene and its product (oxalate oxidase) are already widespread in human food and the environment. Given that oxalate oxidase causes no known harm (beyond its harm to fungal pathogens), it is unlikely to be harmful in the American chestnut genome. But in its new location, might the gene possibly cause some sort of unexpected and harmful interaction? Researchers can test for this kind of harm. For example, Powell and his colleagues have already conducted studies to measure whether the transgenic trees cause harm to bumble bees, other plants, herbivorous insects, frog tadpoles, and symbiotic fungi (different species from the blight fungus). So far, they haven't identified any significant effects.

With no known environmental harm, the next question is whether transgenic chestnuts could affect human health, especially through eating the nuts. The transgenic chestnuts are known to be different from wild ones in only one way: they contain measurable amounts of oxalate oxidase. But because this enzyme is found in many foods that people eat regularly, consuming the nuts would be unlikely to result in any health concerns. Furthermore, in nutritional testing, no difference has been found between wild and transgenic nuts. I asked Powell if he's ever tried one of the nuts from the transgenic trees: "No, I have not. I'm waiting for the regulatory process to say, hey, it's OK. I will be probably the first one to do that. . . . I will . . . give them to my family. There's no problem with those nuts."

Despite efforts to test the safety of transgenic organisms, some people remain unconvinced. For example, the website of the Non-GMO Project states that, "In the absence of credible independent long-term feeding studies, the safety of GMOs is unknown." The organization advocates that those who propose the use of transgenic organisms must first prove that the organisms are safe. This makes intuitive sense, because transgenic organisms should not be available on the market unless they are proven to be safe. However, this kind of argument is inherently open ended. For example, who should decide how long a study needs to last before concluding that something is "safe"?

Anne Petermann, executive director of the Global Justice Ecology Project, has been an outspoken critic of the transgenic American chestnut tree. When I asked her what worried her most about planting these trees, she said, "the risks we don't know about." When I asked her if it was possible to do enough testing to satisfy her concerns, she replied, "You can't do enough. You can't do risk assessments on a tree that lives 200 years in a wild forest ecosystem." She also argued that, even if researchers wanted to do such an impractically long risk assessment, it would be unethical to do so, because the testing itself would enable the transgenic trees to interact with the rest of the ecosystem. "Then, all of a sudden, the genie is out of the bottle," with transgenic pollen, seeds, leaves, and wood interacting with the rest of the ecosystem. If they tried to address Petermann's concerns, advocates for genetically engineered chestnut trees would find themselves in a catch-22: extremely long field tests would be needed to prove the safety of the trees, but such field tests would be considered out of the question because the tests themselves create unacceptable risks.

When I talked to Powell about these concerns, his exasperation was apparent. He said that people want to know "what's going to happen 20, 30, 40, 50 years from now, you know, as if something magical is going to happen to these trees." He said that people worry that the trees are going to "turn into something evil" in the future. "But that's the same argument that you can say for the hybrids of any plant. You don't know what's going to happen to it in the next few decades. So, it's not an argument for genetic engineering [or] against genetic engineering. It's an argument against progress, basically, because it could be applied anywhere."

Some opponents of transgenic organisms argue that the American chestnut is being nefariously exploited as lumber for building a transgenic Trojan Horse. Petermann told me that she believes that by genetically engineering the return of a charismatic tree, the forestry industry is trying to "move the needle on public opinion on genetically engineered trees." She worries that if transgenic American chestnuts are approved for widespread planting, it will open the floodgates for a suite of genetically engineered trees in commercial forestry.

In the end, arguments for or against genetic engineering are often about deeply held differences in values. There's no question that genetic engineering in a lab is an unnatural process, even if it closely mimics a natural one. And even with no evidence that genetically engineered trees cause any harm, they still may be perceived as a frightening fusion of different species into a scientist's new monster. There may always be a question of whether this technology should be applied, no matter the potential benefit; to some, the ends may never justify the means.

Although transgenic organisms have attracted a lot of controversy, hybrid organisms have not. Hybridization and backcrossing of American and Asian species of chestnuts produced new kinds of trees that started with 30,000 copies of each gene from each parent species, which were mixed into new combinations through the backcrossing program. From the perspective of the American chestnut, this means the tree was genetically modified to add potentially 30,000 new gene sequences compared with a single one for the transgenic trees. There's no scientific reason to believe that these new hybrid genetic combinations will be less prone to environmental or human health risks than the transgenic trees. And every time any of these hybrid trees is crossed, it's like "rolling the dice," explained Dr. Jared Westbrook, director of science at The American Chestnut Foundation. "Essentially, every single [hybrid] tree we make has a different combination of Asian and American genes." But unlike the transgenic trees, the hybrids require no testing or approvals for either planting or consumption.

Powell is nevertheless optimistic that attitudes about transgenic organisms are changing. "I'm hoping that the public will eventually come around and say, 'oh, genetic engineering is actually a way to preserve the integrity of species more than all these older techniques,'" such as hybridization, which people have already accepted. Fitzsimmons, who for twenty years has been giving public talks about the efforts to restore the American chestnut, said that she's seeing a shift in attitudes. In the past, she encountered "slightly hostile" reactions to planting transgenic trees, but the people she encounters today are "generally very positive about the technology."

The merit of transgenic American chestnut trees moved from the hypothetical to the practical in January 2020, when Powell and his team petitioned for approval from the US Department of Agriculture to plant one kind of their transgenic trees freely. "We've been working on [USDA approval] for probably over five years," Powell told me, noting that the petition is "280 pages, with a thousand references." The petition has gone through a period of public comment, and the USDA is now taking the next step of writing an environmental impact statement for a second round of open comments. If the USDA approves the plan to plant the trees, Powell will need to get additional sign-offs from the US Food and Drug Administration and the Environmental Protection Agency. As a result, transgenic American chestnut trees could be approved for large-scale planting—and eating—within the next few years.

GENETICALLY ENGINEERING CONSERVATION

Even though the current efforts to genetically engineer a new American chestnut tree do not yield perfect blight resistance, they do offer a glimpse into the future of human-assisted evolutionary rescue. Indeed, although the American chestnut is among the first species to be genetically modified in the name of conservation, it certainly won't be the last. Many other plants and trees are fighting for their lives against a suite of new pathogens that have traveled to new places with the unintentional help of people. In some cases, genetic engineering may be a realistic option for imbuing an organism with the ability to fight new pathogens. As Lisa Thomson noted, many tree species are "under fire from introduced pests and pathogens right now," and "ash, hemlock, elm, you name it," could benefit from similar efforts.

Genetic engineering may also prove helpful in avoiding animal extinctions. For example, genetic engineering has been proposed to help endangered black-footed ferrets battle diseases such as sylvatic plague and canine distemper. And some scientists are trying to genetically engineer corals to help them cope with warming oceans.

Ultimately, the most important part of the American chestnut's story may not be the restoration of the trees themselves. Instead, this story serves as an object lesson in the lengths to which people will go, and the technology that is increasingly available, to give a once-doomed species a chance to persist, while raising larger questions about what to do when nature runs out of tricks to help species evolve to withstand new conditions. Should we wield our increasingly powerful evolutionary toolbox to try to bring them back from the brink? And if so, how far should we go in trying?

CHAPTER 5

A Youth Revolution

I COULD FEEL THE NERVOUS ENERGY on the boat as we were nearing the reef. Excitement is common aboard dive boats, as passengers anticipate the amazing experiences that await them. But this energy was different; we were conservationists and scientists heading to a reef that we suspected was in trouble, so our excitement was overshadowed by foreboding.

Our destination was Cordelia Banks, a group of reefs off Roatán, Honduras. I had visited these reefs many times over the years and knew them to have some of the best coral gardens in the entire Caribbean. Since the 1970s, corals have been dying all over the world, but the Caribbean corals have been especially affected, with most reefs losing most of their coral. Cordelia Banks had been a notable exception, and diving here was almost like time traveling to see the corals of the Caribbean fifty years prior.

As our crew tied the boat to a mooring line, I could see immediately that something was wrong. On a typical day at Cordelia, the mustard and orange colors of the shallow corals can be seen through the clear water. That day, however, they did not look typical; they looked as though someone had shone a black light on them, making them glow in an unnaturally bright neon yellow. This was what we had feared: Cordelia's corals were bleaching.

Bleaching occurs when the corals are stressed, usually because the ocean water is warmer than normal. It was October, and the long, hot summer had not yet yielded to the cooler, rainy weather of fall. As I jumped into the water, I could feel just how unrefreshingly hot it was. My dive gauge showed that the water was a sweltering 90°F (32°C), which was about the same temperature as the air above.

As I dropped below the surface and started Darth Vader breathing through my scuba gear, I could see the scene clearly: entire underwater fields of coral were in the process of shifting from their typical earth tones to abnormally bright yellows and ghostly white. Although coral bleaching is a natural process, it signals that the corals are in acute distress. I felt a knot in my stomach as I contemplated whether this was the beginning of the end of Cordelia's exceptionalism. One thing was certain, however: if the water temperatures didn't drop soon, many of the corals would die.

Scientists, coastal communities, policymakers, and conservationists alike are desperately trying to find ways to save corals before they disappear entirely. As the urgency to save coral reefs grows, people in places such as Roatán are beginning to grapple with ethical questions about whether and how to intervene. Should they rely on natural processes such as evolutionary rescue to save their reefs? Should they try to boost evolution through selective breeding or genetic modification? Should they introduce new coral species from other parts of the world? The good news for coral reefs and the people who depend on them is that there are many ways that nature can rescue corals and a growing number of tools available for people to use to help.

CORAL REEF DIVERSITY

Coral reefs have exceptionally high biological diversity and include more species than any other habitat in the ocean. Our oceans cover three-quarters of the Earth, and although coral reefs occur in only a tiny fraction of this expanse—less than 1 percent of the ocean floor—this sliver of habitat is crammed with roughly 25 percent of all ocean species.

It's one thing to recite statistics about diversity, but it is quite another to experience it firsthand. The first time I experienced a reef, I was overwhelmed by the diversity of life I saw there. I can still remember that first dive like it was yesterday. It was 1991, in Saint Croix in the US Virgin Islands. After descending into the water, I was surrounded by a world of colorful fish swimming through a forest of coral. I saw skinny, bright yellow fish; small fish with electric-blue spots; finger-sized fish that swam upside down, with purple heads and yellow tails; saucer-shaped fish with black and white bands; and beaked fish in pastel pinks and purples. Some corals had delicate, interwoven branches, while others were shaped like giant pillars. Still others were the size of boulders, with brainlike grooves on their surfaces. Auburn barrel sponges were big enough to crawl into. Blue sea fans waved in the ocean's currents. There seemed to be no end to the variety of sizes, shapes, and textures of life on the reef.

After thirty years and thousands of dives, I have never seen anything that rivals coral reefs for their diversity of life. Not only are thousands of species present, but many of them are out in the open, flaunting their colors as they communicate with one another about who would make a good mate, who would be willing to remove whose parasites, who stings, or who's poisonous to eat.

Ensconced within the coral is another world of cryptic species: octopuses that change their colors and textures to mimic their surroundings, crabs that attach sponges to their shells as camouflage, and fish that you'd swear were rocks until they flex their massive jaws to suck in a smaller unsuspecting fish swimming by. The foundation for all this life is the coral. Without the underwater scaffolding corals build, without the habitats that their bodies create for other species, these magical reefs would simply not exist.

CORAL BIOLOGY

Corals are so foreign to humans that many people don't know what to make of them. Are they plants, animals, or something entirely different? I've even heard people describe them as "slimy rocks." Coral are, in fact, animals. If you look closely at a typical coral on a reef, you're likely to see that its surface shows a repeated texture of bumps or pits, often in starlike shapes. Each of these houses an individual coral animal, or polyp, which looks like a tiny sea anemone with a ring of tentacles on top. All the polyps in a given coral are descendants of a single founder polyp that kept growing and dividing to create a colony of genetically identical clones.

Globally, hundreds of species of corals build reefs. Often known as hard corals, they have rocky skeletons, unlike their soft coral relatives with leathery skeletons. (Although both hard and soft corals grow on many reefs, my focus in this chapter is on hard corals, because they are the most threatened and they build reefs. For simplicity, I will hereafter refer to hard corals simply as corals.)

Over millions of years, corals have evolved a suite of incredible bio-logical tools that are at the very heart of how a coral reef works. To under-stand why the reefs are threatened, and how the rescue effect might help them, it helps to understand some of these unusual traits, such as their ability to hunt and farm simultaneously, create rock out of water, and travel from reef to reef.

Corals are hunters and farmers

Like all animals, corals must eat to survive. Because adult corals are anchored to the base of the reef, they can't go searching for food, so they have evolved to get their food delivered in two ways: by hunting and by farming. Corals are ruthless predators that attack and eat small plank-tonic animals that drift by in the water. To dispatch their prey, corals have evolved special stinging cells on their tentacles, which function as poisoned spear guns. When an unsuspecting creature bumps into a hair trigger on the tentacle's surface, the stinging cell fires a barbed spear, puncturing the prey and pumping a dose of immobilizing poison through a tube attached to the polyp. The polyp then reels in the spear and delivers the food to its mouth.

To augment the food they can get from hunting, most corals have taken on a second job: farming. Healthy coral reefs are usually surrounded by nutrient-poor water, making it difficult for algae, the ocean equivalent of plants, to grow quickly. Ingeniously, corals have taken advantage of one of the few places on the reef where nutrients are relatively plentiful to grow a crop of algae: inside their own bodies. Like all animals, coral bodies make metabolic wastes from the food they ingest. Rather than release these nutrient-rich waste products into the water column, corals use them to nourish tiny zooxanthellae algae that live symbiotically within their tissues. These algae benefit by getting a great apartment, with good light and nutrients delivered to their doorstep. In return, they give the coral some of the food they create through photosynthesis.

By combining hunting and farming, corals gain more food energy than they could by just doing one or the other. The corals put this energy to good use by building the reef.

Corals make rock out of water

A thriving reef is like a coral-packed Manhattan, and because there is lots more demand than there are available places to live, many would-be residents are excluded. For people who own space in Manhattan but want more, their best bet is to build upward, and corals do the same. Unlike Manhattanites, however, corals are not governed by building codes and lot lines. So in addition to growing upward, they use a couple more tricks to get more space. First, when neighbors get too close, corals use special stinging cells to assassinate and eat them. Second, corals can grow over their neighbors, shading them from the sunlight they need to grow their own food. My favorite example of this is the tabletop coral, which is shaped like a round table with a single thick pedestal in the center. I've seen tabletop corals that have grown to 10 feet (3 m) across, leaving potential competitors withering under them in the dark.

In a coral colony on a reef, most of the substance of the coral is its inert calcium carbonate (limestone) skeleton, with just a thin veneer of living coral polyps on the surface. Corals make these skeletons by pulling calcium ions and carbonate ions out of the water and chemically fusing them to create calcium carbonate. Because this chemical reaction is energetically expensive, corals benefit by being both farmers and hunters, which helps them accelerate their growth.

In the process of making their rocky skeletons, corals create habitats for other creatures, like trees that create the structure of a forest.

If you've ever visited a coral reef, you probably noticed that corals resemble branching shrubs, spreading plates, or boulders. This three-dimensional reef structure provides shelter for fish, shrimp, snails, crabs, and other creatures. Corals do more than trees can do, however, because they also build the underlying reef, which is analogous to building the hills or mountains on which the forest grows. Over thousands of years, corals have built huge structures, including the Great Barrier Reef in Australia, which, at 1250 miles (2000 km) long, is arguably the largest biologically built structure on the planet.

Corals are ocean voyagers

Like a tree in a forest, a mature coral is firmly anchored in place. But that's not true for baby corals, or larvae, which can travel up to hundreds of miles before settling on a reef.

Living in a colony creates a challenge for coral sexual reproduction. If corals mated like mammals do, males would directly insert sperm into a nearby female. But the immediate neighbors of a given coral polyp are the genetic equivalent of identical siblings. For any organism, whether a tiger or a coral, mating with close relatives is an evolutionary no-no, because the offspring of close relatives often have severe genetic problems. To mate with unrelated individuals, corals release their gametes (eggs and sperm) directly into the water column and hope the right eggs and sperm will be brought together by ocean currents.

For coral sex to produce larvae, different coral colonies need to be ready to release their gametes at the same time. Many corals have therefore evolved to spawn synchronously, in a great watery orgy, by tracking the phases of the moon. The ideal conditions for spawning can look like a plan for a clandestine tryst. For example, in Roatán, many corals spawn in the late summer, about a week after the full moon, and an hour after sunset, before the waning moon has risen. Spawning at night helps corals avoid predators such as fish that would see and happily eat their gametes and embryos in the light of day.

Once the embryos are swept away from the reef, they enter the big, blue, open ocean. Whenever I snorkel in the open ocean, I experience vertigo, because there's no sense of scale or distance or grounding—just endless blue in every direction. The only discernable feature is the water's surface, where air and seawater meet. In this vast environment, the larval corals begin to grow, relying on their transparent bodies to help hide them from predators (imagine trying to find a contact lens in the ocean).

After a period of days to weeks, larval corals settle down to start growing colonies of their own. Depending on the currents, larval corals can settle on the same reef site used by their parents, on a site hundreds of miles away, or on a site anywhere in between. Because larval corals are ocean voyagers, separate reefs in the same part of the ocean are intimately connected by ocean currents.

ROATÁN'S REEFS

The island of Roatán is about 35 miles (56 km) north of mainland Honduras. The largest of Honduras' Bay Islands, it is 30 miles (48 km) long and about 3 miles (5 km) wide, almost banana shaped, and surrounded by coral reefs. In many ways, Roatán epitomizes the challenges of saving coral reefs, which are simultaneously highly valuable and increasingly at risk.

A valuable asset

Roatán is part of a developing country that is seeking economic opportunities for its growing human population. The island's reefs are some of its most valuable assets, because they draw visitors from around the world who come to snorkel and scuba dive, providing an infusion of international cash into the local economy. Hotels, restaurants, taxis, and tour operators alike benefit economically from the reefs.

Roatán also has a large group of artisanal fishers who look to the ocean to feed their families or sell their catch to hotels and other buyers. Some are professionals with motorboats, who travel around the islands, while others are amateurs trying to catch something for dinner.

Years ago on a Roatán reef, I encountered a fisherman whose only gear was a patched dive mask and a spool of line with a baited hook. He swam over the reef, dangling his hook, trying to encourage anything to bite. He was so desperate to catch food to bring home that he'd kept juvenile fish that were no larger than potato chips. Like many other families in the area, his family's nutrition was closely connected to the health of the reef.

The full economic value of the island's reefs extends well beyond local dive tourism and fishing, however. Roatán's barrier reef protects shorelines from damaging storms, and its beautiful white sand beaches are created from broken bits of coral.

One study, for example, estimated the value of the island's natural environment, which is dominated by the ocean, at $1.3 billion. Moreover, Roatán is not alone in relying on its reefs for economic opportunities; by one estimate, coastal communities comprising half a billion people rely on coral reefs for food and economic opportunities globally. Furthermore, coral reefs house a share of the world's medicine chest of the future. Many new medicines are based on chemicals produced by living organisms, and coral reefs are already the source of potential new pain killers, cancer drugs, and antivirals. Finally, thousands of other species rely on the corals to build their homes. Some species could adapt to a world without coral, but others are so reliant on them that, without corals, they'd disappear. With so much value at stake, communities in Roatán and around the world are strongly incentivized to save their reefs, but to do that, they need to mitigate a growing list of threats.

The struggle to keep up

Corals started to decline on Caribbean reefs in the 1970s. At that time, live corals covered about 50 percent of Caribbean reefs. By the year 2000, the cover of corals in the Caribbean had declined by 80 percent, to an average of about 10 percent live cover. For comparison, when a forest loses 80 percent of its tree cover, it's no longer considered a forest; it's a savanna.

Roatán's reefs have fared better than most other reefs in the Caribbean. According to the Healthy Reefs Initiative, which monitors coral reef health in the region, a typical Roatán reef included about 30 percent live coral cover in 2018. Although this is much better than other parts of the Caribbean at the time, the 2018 percentage probably represents half the amount of coral that inhabited the reefs in the 1970s, and there are concerns that the coral cover will continue to wane.

Paradoxically, the value of the island's reefs may be partly to blame for their decline. Consider the community of West End, whose shore is home to a long ribbon of coral reef that is protected from the large waves caused by the northeast trade winds. The local dive industry realized many years ago that an alluring, wind-protected reef was a recipe for economic success. Accordingly, West End has some of Roatán's most economically valuable reefs and has become the center of the island's dive tourism, which has resulted in increases in fishing and pollution.

Historically, West End's reefs have been open to extensive fishing using nets, traps, spears, and hook and line. Coral reefs can handle some level of fishing without experiencing too many adverse effects. However, as the tourism industry grew in the 1980s and 1990s, so did the local population of fishers. In general, as fishing pressure increases, the most prized fish start to disappear, which can cause the whole ecosystem to shift. One group of fishes, the grazers, benefit corals by mowing down seaweeds, keeping space open for corals to grow. When grazers are fished out, seaweed can grow more freely and may take over space once occupied by corals.

As fishing pressure was increasing, the growing community of West End was also creating sewage that was not treated for many years and was directly released underground or into irregularly maintained septic systems. From there, it would swiftly flow into the ocean and the reef. Beyond the "ick factor" and risk of spreading waterborne disease, untreated sewage brought an enormous influx of nutrients, which was bad news for corals, because high nutrient availably favors fast-growing seaweeds over slow-growing corals. In sum, the one-two punch of low fish grazing rates and high nutrients helped fuel an explosion of seaweed, while corals declined.

In addition to the coral declines caused by fishing and pollution, Caribbean reefs have suffered from wave after wave of coral diseases since the 1970s. Most recently, reefs have been suffering from stony coral tissue loss disease, a mysterious, fairly recent disease that attacks twenty-four species of stony corals (hard corals, the type that build reefs), often killing entire colonies within weeks or months. Scientists have not yet identified what causes the quickly spreading disease. Locals first noticed it in Roatán in September 2020, and two months later, it was already spreading around the island like a storm front, racing from one reef to the next.

The reefs in West End are also experiencing their biggest threat of all: coral bleaching. As the Earth's climate warms, as much as 90 percent of our extra heat has been absorbed by the ocean. By acting as the planet's air conditioner, the ocean has provided a huge service to life on land, but the downside of this service is that the ocean is rapidly warming.

Ocean water of just a few degrees above normal can trigger coral bleaching. Although the exact cause of bleaching is still being studied, one leading hypothesis starts with the observation that, over millennia, corals have fine-tuned their relationships with symbiotic algae to achieve an optimal growth rate, favoring strains of algae that perform the best under typical environmental conditions. When the water becomes unusually hot, the algae have trouble managing photosynthesis, because their biological machinery is geared to work at lower temperatures. (Imagine baking a loaf of bread in an oven that is too hot—the final product probably wouldn't be what you'd expected.)

When photosynthesis is impaired, the algae can create new sets of incomplete molecules, some of which are highly reactive and unstable free radicals that can chemically react with important molecules in their vicinity, undermining key biological functions within the coral. For humans, dietitians sometimes recommend that we consume antioxidants, such as those found in blueberries or red wine, because they can neutralize free radicals. As ocean temperatures rise, corals cannot produce enough antioxidants to neutralize the free radicals that are becoming increasingly toxic. In a last-ditch effort to reduce the production of more free radicals, the corals expel some or all of the algae.

Because coral animals are typically transparent, the color of a coral colony on the reef is mostly derived from the symbiotic algae. As some algae are ejected, coral colonies begin to grow pale and can appear to glow. Bleaching occurs when all the coral's algae are gone, and the bright white calcium carbonate skeletons are visible through the still-living coral animals. At this point, the already sick corals begin to starve without their crop of algae. If the water stays warm for too long, stressing the coral and preventing the regrowth of algae, the corals will die.

Prior to 1980, coral bleaching was observed as a rare, localized event. The first global bleaching event occurred in 1982–83, as tropical waters around the world grew especially warm. Since then, coral bleaching has progressively become more common, with nearly every reef on the planet having bleached at least once in the last forty years. Roatán's reefs have bleached at least nine times since 1995 and bleaching is now an almost annual event. When bleaching is added to the other threats harming corals today—such as pollution, overfishing, and disease—it's no wonder that Roatán's reefs are struggling to stay alive.

Community management

Residents of Roatán first noticed problems on their reefs in the 1980s. At that time, dive guide Tino Monterroso was working in Sandy Bay, about 2 miles (3 km) up the coast from West End. He told me that he was concerned that the fish were disappearing because of too much fishing pressure. A fisherman himself, he convinced other local fishers that creating a marine reserve, where a portion of the reef was closed to certain forms of fishing (such as spears or nets), would improve their catch in the long run. Inside the reserve, protected fish would act as brood stock to seed other reefs with more newly hatched fish. These protections would also be good for tourism—divers like to see big, charismatic animals such as groupers, barracuda, and sea turtles that may otherwise be over-fished in legal and illegal fisheries. The community agreed to establish a marine reserve in Sandy Bay, and a few years later it was expanded to include West End.

Honduras is a developing country that faces extraordinary challenges, including poverty, corruption, and drug trafficking. Understandably, park management is not always at the top of the list of government priorities. The local community knew that if they wanted their marine reserve to work, they would have to take the lead on managing it. Enforcement in marine reserves is expensive and challenging, but the community was undeterred. To finance management, they asked all dive operations in the area to charge their customers a minor and voluntary reef preservation fee to fund actions such as enforcement. Eventually, the community even established a nonprofit organization, the Roatán Marine Park, which is dedicated to reef management.

Building up enforcement capacity was a slow process. According to several of the locals that I interviewed, it wasn't until the mid-2000s that West End's reefs were regularly and effectively patrolled. As a result, the numbers of fish on these reefs have begun to grow. Based on data from the Healthy Reefs Initiative, between 2006 and 2016, the abundance of grazing fish doubled in the reserve. The same results were noted for big fish such as groupers and snappers, which are most prized by fishers.

These days, you don't need to see the data to notice the difference between West End reefs and those in other parts of Roatán. A few years ago, as the executive director of the Coral Reef Alliance, I hosted a group of divers on a trip to see the island's reefs. We spent several days diving in various areas around the island but saved West End for last. Upon surfacing after our first West End dive, the divers couldn't get over the schools of huge jacks and snapper, the giant parrotfish (an important grazer), and the abundant turtles. The reefs near West End were unlike anything else they had seen on other reefs around the island. Community-based enforcement was clearly working.

As the fish populations were recovering on West End's reefs, the community turned to the next challenge: cleaning up its coastal waters. The biggest improvements came after a new wastewater treatment plant was built in 2011. Since then, businesses and homes have been slowly connecting to the sewage system, and today almost all the community's sewage is treated. It's easy to see the difference this has made to the ocean environment: for example, the waters of Half Moon Bay, in the heart of West End, are now visibly clearer than they were just a few years ago.

Other Honduran communities have noticed the success of these endeavors in West End and have been inspired to try similar tactics to help protect their reefs. For example, Cordelia Banks, the reef with unusually abundant stands of corals, was declared a marine reserve in 2012, and today fishing is forbidden on its shallow reefs, which has enabled fish populations to start rebounding. At the same time, the nearby community of Coxen Hole is working to reduce coastal pollution by ramping up the capacity of their own wastewater treatment plant. The fisheries management patrols that started in Sandy Bay and West End have been extended to a patchwork of newly protected reefs that ring the entire island of Roatán. Similar community-driven reef conservation projects are popping up on neighboring islands Utila and Guanaja and on the mainland.

So far, local management seems to have been enough for West End's corals to hold their own. The Healthy Reefs Initiative has been surveying reefs in the vicinity of West End since 2006, when the local reefs had 24 percent live coral cover. In 2016, live coral cover remained at 24 percent. By 2018, after bleaching occurred in the three previous years, live coral cover rose slightly, to 28 percent. As reefs around the world are experiencing declining numbers of live corals, the fact that these reefs are holding steady is something to celebrate.

NATURAL RESCUE

Nature should have a better chance of saving West End's corals because of the successful management practices instituted by the local community. In particular, because of efforts to protect grazing fish and ensure clean water, phenotypic, reproductive, demographic, and evolutionary rescues are all likely to function better.

Phenotypic rescue

Corals can survive across a range of environmental conditions by adjusting their physiology, appearance, or behavior through phenotypic rescue. However, as with the mountain pygmy-possums, the more stresses simultaneously encountered by a coral, the less likely it will be able to adjust successfully. Reducing chronic stressors such as water pollution or competition with seaweed may give local corals more phenotypic space to deal with harder to control stressors such as disease epidemics and bleaching.

Jenny Myton, conservation program director at the Coral Reef Alliance, told me that she's seeing evidence of phenotypic rescue in West End. She has been involved in almost every major reef conservation action in Roatán for the past fifteen years, including West End's improvements in fisheries management and water treatment. She and her colleagues have recently noticed that incidences of coral bleaching and disease have been declining in places where water quality has improved the most.

Corals can also deal with a stressful situation more successfully if they've encountered it before. For example, if corals can survive bleaching caused by warmer water temperatures, they may shift their phenotype to cope better when again confronted with more warm temperatures. One way that corals can do this is by bookmarking the genes that they used to survive bleaching, so that when the water heats up, their bodies will know exactly which genes to turn on to survive. The first bleaching event is a bit like a fire drill: the corals "practice" what to do in an emergency, so that a phenotypic response can be more efficient and effective the next time. Corals that survive bleaching can also change their phenotype by taking in different strains of symbiotic algae that perform better under warmer temperatures, like an elite sports team trading for new players to fine-tune performance.

Reproductive and demographic rescue

The work that the West End community has done to protect grazing fish and clean up water pollution should also increase the chances for both reproductive and demographic rescue. For example, if more corals survive on the reef through phenotypic rescue, they are likely to produce more offspring. Though some of these babies will drift far away, others will stay close to home, increasing local coral populations through reproductive rescue.

Reefs with clean water and high rates of fish grazing are known to provide better habitat for young corals, so they may be more likely to survive to adulthood. This is true for the larval corals that return to the reefs where their parents reside (reproductive rescue) as well as corals that drift to West End from nearby reefs (demographic rescue). As a result, local management of reefs should result in an increase in the abundance and survival of the next generation of corals.

Evolutionary rescue

Although phenotypic, reproductive, and demographic rescues will help sustain coral populations in the short term, eventually, as climate change intensifies, conditions will likely become too harsh for local corals, and nearly all of them will die. To ensure the survival of future generations, Roatán's corals must undergo evolutionary rescue, whereby they evolve new traits to cope with adverse environmental changes. Fortunately, we know that corals can evolve to deal with warmer ocean temperatures. The oceans in the Middle East, for example, are some of the warmest in the world. Over millennia, corals there have evolved remarkable tolerance for the heat. Given enough time, Roatán's corals may do the same—but can they evolve quickly enough to make a difference?

As was the case with the American chestnut, a key ingredient for rapid evolutionary rescue is heritable genetic variance, which is a measure of how much genetic variety exists in a population for a given trait, such as disease resistance or thermal tolerance. Generally, a population with more genetic variance is more likely to include some individuals that can tolerate changing environmental conditions. American chestnuts did not have enough genetic variance for evolutionary rescue to save them from the chestnut blight; however, corals may possess enough variability to enable them to keep up with changing ocean conditions.

Some corals, in fact, have been shown to be genetically predisposed to deal with high temperatures, pollution, or other stressful conditions. These *super corals* are so named because of their unusual ability to survive environmental extremes. (Scientists also use the term *extremophiles* to refer to organisms that thrive in conditions that would be too harsh for others.) Evolutionary rescue could theoretically occur quickly if super corals disproportionately survive under harsh conditions and seed reefs with a new generation of super corals. In fact, researchers have uncovered circumstantial evidence that indicates this is already happening. In a global analysis of coral bleaching from 2007 to 2017, researchers discovered that the temperature threshold for coral bleaching has increased by about 0.5°C. It's likely that this temperature tolerance is at least partly a result of evolution (in this case, probably the disproportionate survival of tolerant corals) and partly a result of phenotypic rescue.

Adaptation networks

While I was the executive director of the Coral Reef Alliance, I worked with a group of researchers to try to determine whether a combination of phenotypic, reproductive, demographic, and evolutionary rescue would be strong enough to help corals keep pace with climate warming in the future. We used mathematical models to explore the possible fate of corals through 100 years of climate warming, followed by 500 years during which the climate stabilized at higher average temperatures.

The first results, published in 2019, were cause for cautious optimism. The modeled corals could adapt quickly enough to survive climate change. To reach this result, evolutionary rescue would be required, and the higher the genetic variance the better. Corals also fared better if they were part of an adaptation network, an array of well-managed reefs connected to one another across a region. In other words, similar to how connected and well-managed parks are helping tigers in India, connected and well-managed reefs can help rescue corals. With corals, connectivity within networks is relatively easy to establish. Because coral larvae disperse through open water, no protected corridors are needed; protected reefs must simply be close enough to one another for larvae to drift from one to the next.

West End's corals live within the Mesoamerican Reef region, which spans the Caribbean coasts of Mexico, Belize, Guatemala, and Honduras. The larval corals that settle in West End may travel from faraway reefs located along the shores of any of these countries, and these populations are likely to include some super coral offspring.

West End is not alone in taking action to protect local reefs. Similar protections are being enacted throughout Roatán, on neighboring islands, along the Honduran mainland, and in neighboring countries.

Collectively, this network of well-managed reefs can fuel the rescue effect with a combination of phenotypic rescue (higher survival resulting from good local management), reproductive rescue (greater local reproduction), demographic rescue (lots of larvae arriving from elsewhere), and evolutionary rescue (higher survival of stress-tolerant individuals).

Although these results are encouraging, our study also predicted dark days ahead. Many corals will die as climate change continues to cause rising ocean temperatures. Indeed, this death is critical for evolutionary rescue to work: the flip side of the survival of the fittest is the death of the less fit. The study also made a critically important assumption: that humans will figure out how to stop ongoing climate change during the next 100 years. If we don't do that, these cautiously optimistic conclusions are probably unwarranted.

BOOSTING THE RESCUE EFFECT

I asked Jenny Myton, "With all the progress in places like West End and the potential for coral adaptation, are you still worried about the reefs in Honduras?" She answered swiftly, "Every day. All the time. It's at the bottom of my belly, for sure." While local interventions are demonstrably helping, corals are far from out of the woods.

A key question for the future is whether the rescue effect will ultimately be enough to save corals on its own. Although research may offer some reason for hope, not everyone is content to leave the fate of coral reefs up to nature alone. Some people are actively looking for ways to boost the rescue effect. To date, most efforts have focused on boosting either reproductive or evolutionary rescue using corals grown in captivity.

Boosting reproductive rescue

The most straightforward way to boost the rescue effect for corals is through reproductive rescue. In coral conservation circles, this is known as restoration and it involves directly manipulating the birth rate, growth, and survival of individual corals. The theory is simple: corals are dying, so we must replace them with more corals.

The simplest form of coral restoration is coral gardening, a process that is similar to how a gardener divides and replants tulip bulbs that had grown into a dense clump. Scientists start by dividing coral colonies into small pieces. These broken pieces include intact polyps, and they can each be grown into separate colonies under controlled conditions to reduce mortality. When the colonies reach a desired size, they can be transplanted back on the reef. And voilà! A coral garden!

But there is a problem with simple restoration: if the issues that caused many corals to die in the first place are not addressed, the newly transplanted corals will suffer the same fates that befell their predecessors. Ideally, this restoration effort will be paired with efforts to reduce overall stress on reefs, and this is what several groups are already doing by transplanting hundreds of corals to protected reefs in Roatán each year.

Restoration efforts may prove successful for small-scale projects to boost a particular reef for local benefits. The challenge for restoration of coral reefs more broadly, however, is the cost of scaling it up: coral reefs may be only a small part of the world's oceans, but they are large enough to support billions of coral colonies. A 2019 scientific study estimated the cost of restoring reefs through coral gardening at $351,661 per hectare. Roatán alone has thousands of hectares of reefs, but these are just a drop in the bucket globally. By one estimate, the world has 96,500 square miles (250,000 square km) of coral reef habitat. Restoring all of that would result in a jaw-dropping price tag of $9 trillion.

Boosting evolutionary rescue with super corals

Coral restoration may become more promising if it is combined with deliberate efforts to speed up evolutionary rescue. Perhaps the most straightforward way to do this is to focus restoration efforts on super corals. Scientists are identifying some super corals that occur naturally on reefs. When no naturally occurring super corals are available, however, corals can be bred in tanks, where breeders can artificially select for traits such as thermal tolerance. Research is also underway to try to genetically engineer more resistant coral strains. Once super coral populations are identified and/or created, conservationists may be able to propagate them in large numbers to be transplanted on reefs. If the transplanted super corals survive better than their wild peers, they may disproportionately contribute to the next generation of corals, accelerating evolutionary rescue.

Though efforts to boost coral populations through evolutionary rescue are conceptually promising, scientists don't yet know whether they will be practically feasible. After all, it would take a very large number of transplanted super corals to accelerate evolution. Coral reef regions can have millions, or billions, of coral colonies; trying to boost their evolution by adding a few super corals would be like trying to turn a bucket of white paint (wild corals) into red paint by adding a single drop of red paint (super coral transplants) at a time. This dilution problem is further exacerbated by the fact that climate change is a moving target. The super corals that survive today may not survive the climate challenges of the future. As the corals begin to evolve, their environment may again shift enough to cause even super corals to die.

I have been involved in a recent modeling study to estimate the amount of effort needed to boost the evolution of corals on a network of coral reefs, such as the Mesoamerican Reef. Our model shows that restoring super corals could indeed help, but only when the effort continues at high levels every year for a century. Unfortunately, anything short of that shows little benefit. Thus, boosting evolutionary rescue for corals may require commitment to a large, long-term investment to achieve significant positive results.

Boosting genetic variance for evolutionary rescue

One challenge to using transplanted super corals to supercharge evolution is that the practice assumes that scientists will be successful at guessing which traits corals need in order to survive in a changing environment. It's easy to conclude that traits such as temperature tolerance will be important, but how will these temperature-tolerant corals respond to other stressors, such as disease? Super corals may be successful at surviving one stressor but may fail to survive another.

I once watched a race between two world-class athletes: Olympic sprinter Carl Lewis and champion boxer Evander Holyfield. In a 100 meter race, Lewis was once considered the fastest person in the world, but this race was 800 meters, well beyond Lewis's narrow sweet spot. In contrast, boxers like Holyfield train to strengthen several traits—they need to be fast, but they also need strength and endurance. Predictably, Lewis began the race in the lead, as he ran faster than Holyfield, but he slowed down as the length of the race exceeded his sweet spot. Holyfield eventually caught up to Lewis, passed him, and won the race. Similarly, if conservation advocates single-mindedly breed temperature-tolerant super corals, they run the risk of creating limited specialists. To succeed, future reefs may need elite generalist corals that can handle a wide range of new conditions, including a combination of increasing storms, disease, pollution, and temperatures.

Roatán's reefs are experiencing the need for elite generalists first-hand, with the arrival of stony coral tissue loss disease. Myton told me that some reefs have already lost half of their coral to the disease. But she and her colleagues have noticed that some corals seem immune and are surviving where others have died. The survival of these corals offers hope that corals may already be evolving to keep pace with their changing environment. It also underscores that although the corals that succeed in the future will need to survive warmer ocean temperatures, they'll also need to be able to cope with many other stressors.

Rather than putting all their eggs in the temperature-tolerant super coral basket, scientists may see more successful restoration efforts if they focus on increasing the overall genetic variance of corals, while letting nature choose the winning traits. For example, perhaps corals of the same species that hail from distant locations such as Brazil or the Bahamas could be brought to Roatán. Distant populations are almost always genetically different from one another, so this could provide more genetic variance to Roatán reefs—more tickets to the environmental change lottery. Within the transplanted population, a few corals may have the perfect combination of genes to thrive in Roatán's changing waters.

Dr. Mary Hagedorn, a senior research scientist at the Smithsonian Conservation Biology Institute and the Hawai`i Institute of Marine Biology, is experimenting with a new way to increase local genetic variance by creating a sperm bank for coral. In the wild, corals from distant reefs could never mate, because their gametes would never meet. Hagedorn is hoping to solve that problem by cryopreserving (freezing) coral sperm collected from around the world to be used in the future.

As a proof of concept, Hagedorn and her colleagues have made crosses of elkhorn coral using eggs from Curaçao and cryopreserved sperm from Florida, Curaçao, and Puerto Rico. These three locations are hundreds of miles apart from one another, and each local elkhorn coral population is likely to be genetically different from the others. Hagedorn's crosses have produced hundreds of young elkhorn corals, which were growing in tanks in Florida when I spoke with her in 2020. Many of the young corals have totally new combinations of genes, which Hagedorn described as "an extraordinary treasure trove of diverse genetics that could completely diversify the Florida reefs that are struggling to maintain themselves." However, she said with some exasperation, "no one will put them back out on the reef." When I asked her why, she said she was unsure of the exact reason but conceded that "sometimes, ideas are ahead of their time and people are fearful of taking the risk to embrace them." She asked rhetorically, "What's the point of . . . technological interventions if we are too afraid to use them?" Perhaps someday conservation managers, and the laws and regulations that govern their actions, will embrace newer strategies such as using cryopreservation to increase local genetic variance. In the meantime, these corals may never see a reef, because, as Hagedorn was resigned to admit, "We're running out of money, and people are going to start chucking them."

Although there are important questions regarding whether coral evolution can be boosted across large areas of reef, adding super corals or increasing genetic variance may still provide benefits locally. For example, to ensure that temperature-tolerant corals can persist on key reefs such as those at West End, reef managers may choose to transplant particular corals that they believe will have the best chance of surviving. The net result could be small-scale designer gardens of coral, the underwater equivalent of a botanical park.

Introducing new coral species

As conditions on coral reefs continue to deteriorate, reef managers could try the conservation equivalent of football's Hail Mary pass (a last ditch effort): they could introduce new species of corals collected from far-away reefs to replace species that are declining locally. For example, hundreds of species of coral inhabit the Indian and Pacific oceans, and some are adapted to warmer temperatures. Perhaps some of these corals could survive under current or future conditions in the Caribbean. Indian and Pacific ocean corals would need to be transported quite a distance by humans to the Caribbean. This wouldn't exactly be human-assisted geographic rescue, because its purpose wouldn't necessarily be to rescue the new species from poor conditions in their native range. Instead, it would be considered a *functional introduction*, whereby new species are introduced to a reef in an effort to compensate for those that have already declined (similar to the idea of planting Asian chestnut trees to replace American chestnut trees).

While researching this chapter, I asked many scientists, managers, and conservationists whether new coral species should be introduced to the Caribbean to replace the species in decline. Some were horrified by the idea, noting cases of introductions gone wrong with other species in other places. Others were much more pragmatic. One scientist said, "I want these systems to survive on their own. But . . . if reefs are degrading so much and are not able to rescue themselves with the current mix of species," introducing new species may be the "next best" option.

One of the most significant problems with deliberately introducing new species to an area is that once the new organisms are in place, they cannot necessarily be contained. This is especially true in the ocean, where ocean-voyaging coral larvae would make it impossible to control the spread of introduced species. It's understandable that some people are concerned that a new coral species, once introduced, could cause harm by outcompeting natives. But is this really likely? A key problem in the Caribbean is that corals no longer inhabit much of the space they once occupied, and these spaces are now occupied by species other than corals. As a result, there's plenty of room for new coral species to move in without needing to compete with native corals. Furthermore, one of the hallmarks of a healthy coral reef is the coexistence of many species.

I traveled to Fiji in the South Pacific in early 2020. The local reefs near Savusavu, where I was staying, had been decimated in 2016 by Tropical Cyclone Winston. Four years later, these reefs were in full-blown demographic rescue, after thousands of corals drifted here from distant reefs and settled on the shallow Savusavu reefs, forming young colonies. These colonies included dozens of coral species; in fact, it was difficult to find two that looked the same. If all these different species can coexist on the same reef in Fiji, would Caribbean corals be harmed if a few corals from elsewhere were introduced?

Some people are concerned that introduced corals could thrive to such an extent that they would populate local reefs and spread throughout the region. This is exactly what happened with the orange cup coral, a beautiful little coral with yellow or orange tentacles that usually unfurl at night. Cup corals are a bit unusual in that they lack algal crops and get all their food from predation. Because they don't need to live in the sun to raise algae, they can happily colonize shady ocean habitats, such as vertical walls, overhangs, and caves.

Orange cup corals from the Indo-Pacific were first observed on reefs in the Caribbean in the 1940s. These corals probably made their way to Caribbean reefs by attaching to the hulls of ships, or perhaps their larvae hitched a ride in ship ballast water. They have since spread throughout the Caribbean, but it's unclear whether or not they are causing significant harm. Scientific studies show that cup corals do indeed compete for space with native corals, and this potentially harms their nearest native neighbors. The same could be said for all corals, however, whether native or nonnative, because all corals defend their space.

One of the biggest risks of adding new coral species is inadvertently introducing new diseases, such as stony coral tissue loss disease, which may hitch a ride on the coral transplants (as chestnut blight did on introduced Asian chestnut trees). The risk of introducing pathogens could be reduced by quarantining new corals or introducing coral strains that are sexually propagated in laboratory conditions that aim to separate parent pathogens from their offspring. It's worth noting, however, that species from other oceans may already have evolved resistance to diseases such as stony coral tissue loss disease, just as Chinese chestnuts are resistant to chestnut blight; so introduced corals may be especially well suited to thrive on epidemiologically challenged Caribbean reefs.

Successfully introducing new coral species to a place like Roatán would mean that the composition of the local species would permanently change. A key question going forward is, what do people value more: preserving a historical mix of species, or having reefs that continue to grow and thrive? If the latter is deemed most important, then purposeful introductions are worth careful consideration in places where local corals have declined to such an extent that they no longer provide this value.

The world that we are creating today includes new combinations of species everywhere, as a result of human activity. Although people may value species combinations that existed in the past, ecosystems are quite adept at adjusting to new species. And so are people. For example, I've never heard anybody complain about Roatán's coconut or mango trees, which were transplanted from other parts of the world. Instead, these trees have become part of our collective expectations for a Caribbean island. Perhaps it would be the same for new corals, which could potentially enrich the diversity of life on Roatán's reefs, while building reefs that provide benefits to local species and people.

THE FUTURE OF CORAL REEFS

For a long time, coral reef conservationists have believed that if they do a good job of managing local coral reefs, the reefs can return to what they once were, with the seafloor again covered with large stands of coral that house diverse and abundant fishes and other sea creatures. A key part of this story is that reefs should look more like they did in the 1970s, before corals on many reefs started to decline. The imagined corals on these reefs are like trees in old-growth forests, where centuries of minimal environmental change have enabled a few lucky individuals to live for many human lifetimes and grow to enormous sizes.

As conditions continue to deteriorate for corals, most conservation practitioners have accepted that this story is a fantasy. Many of the old-growth corals that were building reefs in the 1970s are already dead, and most large corals that are still hanging on will be gone in a few decades. And new generations of coral are not living long enough to grow very large.

But that's OK, because we don't need old-growth corals to save coral reefs. In fact, we need just the opposite: a youth revolution that fuels evolutionary rescue. Long-lived corals probably won't help in this process, because they slow down coral evolution. Instead, the kinds of coral that are most likely to win are those that can grow from larvae to a sexually reproducing colony within a decade or two. The most hopeful sign that corals are adapting will be reefs covered in small, young corals that return following the now-inevitable waves of coral death.

The reefs of Roatán demonstrate a promising model of what can be done locally to buy coral reefs more time while this process plays out. Moreover, establishing adaptation networks of well-managed reefs, which promote phenotypic, reproductive, demographic, and evolutionary rescue, may collectively produce a strong enough rescue effect to save corals from extinction. At the same time, people increasingly have the power to boost evolutionary rescue directly, while also considering introducing new coral species to fill the ecological roles of those that have become less abundant.

Attitudes are changing in Roatán about whether and how to boost the rescue effects for corals. Myton told me that in the early 2000s, many people considered any direct manipulation to the corals categorically off limits. But in the last twenty years, with continued threats to coral reefs, some of the most vocal opponents are now the cheerleaders for coral restoration projects. Urgency is the reason for these shifts in attitude. As Myton put it, "We've got to do something now" before all of Roatán's coral is lost. She said that the reefs have been so degraded that more and more people are considering intervening in new ways. After all, what more do we have to lose?

Nevertheless, Myton thinks that Roatán is not necessarily ready for the most extreme interventions, especially importing corals from other reefs, whether they are corals of the same species within the Caribbean or different corals from the Indian or Pacific oceans. She doesn't dismiss these ideas out of hand. Given the growing desperation to save the corals, she said, "Who knows? Maybe in five years we'll have a different opinion."

Indeed, some Roatán residents are already ready to try, including Ian Drysdale, the Honduras coordinator for the Healthy Reefs Initiative. Drysdale has been leading numerous reef conservation efforts in Honduras for fifteen years. When I asked him whether Roatán should consider going to extremes to save corals, he said, "If we're going to have a reef that's alive and we're going to have some coral species, heck yeah, go for it. Do crossbreeding, do nurseries, do GMOs, create Frankensteins. If there's anything we can bring back, I say it's worth exploring." When I asked him whether his enthusiasm extends to bringing in new species, he answered, "If it's going to ensure that Honduras still has coral reefs that still produce fisheries [and] that still sustain livelihoods of the local populations, yes."

Corals are racing to adapt to a changing world. So far, Roatán's corals are managing to keep up. Fortunately, as the urgency to save coral reefs rises, communities like those on Roatán have a rapidly growing list of options to try to help them survive.

CHAPTER 6

A Flock of Cichlids

IN APRIL 1985, while sitting in a sinking boat in Africa's Lake Victoria, Tijs Goldschmidt had an epiphany. The wooden boat was old and leaky, and Goldschmidt's research assistants were frantically bailing water and filling the holes in the boat with cotton to keep it afloat. But Goldschmidt wasn't focused on the state of the boat; his mind was on the fish in the bucket in front of him. For years, he had been studying one of the greatest evolutionary events in history that gave rise to hundreds of species of cichlid (pronounced *sick*-lid) fish in Lake Victoria. Goldschmidt and his assistants had been methodically collecting and describing the cichlids, trying to understand how so many different species had come to be. Meanwhile, in the water below his rickety boat, the ecosystem was undergoing profound changes.

The researchers had just pulled in their net and placed their catch in a bucket to be identified and counted. Based on his years of sampling,

Goldschmidt could see that the fish in the bucket weren't right for this part of the lake. For starters, there were abnormally few fish in the catch. Moreover, the usual species weren't present, while other species usually found in different parts of the lake had turned up. Staring into the bucket, he experienced "a feeling of disorientation, of bewilderment," as though his mental frame, which was so focused on how these fish had evolved in the past, was colliding with the events of the here and now. Then it occurred to him: this bucket of fish represented a small window into an enormous ecological chain reaction that revolved around one particular species, the giant Nile perch, which had been introduced deliberately into Lake Victoria in the 1950s and was now gorging on cichlids.

A MIRACLE IN THE RIFT VALLEY

The African continent is literally being torn in two. Most of us know that the movement of the Earth's tectonic plates creates mountain ranges and triggers earthquakes where they collide. In some places, however, instead of colliding, the plates are pulling apart from each other, creating fissures in the Earth called *rifts*. Perhaps the most famous rift zone in the world is in eastern Africa, where the Somali Plate, which includes parts of Kenya, Tanzania, Somalia, and several other African countries, is slowly spreading southeastward, away from the African Plate, which is moving northwestward. This has triggered numerous volcanic eruptions that released mineral-rich ash that enriched nearby soils, giving rise to incredibly productive grasslands and abundant wildlife in such places as the Serengeti in Tanzania. Some of the great fissures in the surface of the continent have, over millennia, filled with water to create some of the largest and deepest lakes in the world, including Lake Tanganyika and Lake Malawi.

Sandwiched in the middle of all this tectonic action is a broad, relatively flat piece of the Earth's crust that is centered near the modern intersection of Uganda, Kenya, and Tanzania. For millennia, this area served as the headwaters for rivers that flowed westward across the African continent and into the Atlantic Ocean. However, about 400,000 years ago, tectonic movements caused the land in the west to tilt upward, blocking the westward flow of rivers. With no outlet to the west, these rivers pooled in areas with the lowest elevation, forming what is now Lake Victoria. Eventually, the lake filled with so much water that it overflowed. But the water didn't flow westward; instead, it followed a winding path northward, where it became part of the Nile River, which then flowed more than 1900 miles (3000 km) farther before reaching the Mediterranean Sea. Since then, Lake Victoria has oscillated between being a large lake when rains were plentiful and shrinking and drying up during periods of extended droughts.

As Lake Victoria has experienced its literal ups and downs, something of a biological miracle occurred when two species of haplochromine cichlid fish had a chance encounter. Cichlids are a family of mostly freshwater fishes that are predominantly found in Africa and the Americas. The haplochromines are a branch of the cichlid family tree and exist almost exclusively in Africa. They are typically hand-sized or smaller, and many species are strikingly patterned and colored. Dr. Joana Isabel Meier, a research fellow at Cambridge University, who uses genetics to study the evolution of haplochromine cichlids in Lake Victoria, told me that roughly 150,000 years ago, "there was a period of high rainfall," which "led to connections between the Congo and the Nile drainage system." Based on genetic evidence, Meier has concluded that two related species of cichlids came into contact after millions of years of separation, mated, and produced hybrid offspring.

In nature, separate species often do not mate, because hybridizing with another species can be a waste of energy. Hybrid embryos often don't survive early development, and even if they do, the resulting organisms can be sterile or poorly adapted to their environment. Against the odds, this particular cichlid hybrid managed to survive and reproduce, eventually becoming an entirely new species of cichlid in the vicinity of Lake Victoria. From a genetic perspective, this new organism was something of an anomaly, because it was genetically a fusion of two different parent species. On the tree of life, two separate species branches had fused together into one.

Meier believes that the new hybrid species began to take advantage of the opportunities presented by the abundance of water, probably evolving into multiple species. However, the good times for this new group of haplochromine cichlids in and around Lake Victoria came to an end during a climate shift. This time, as rains decreased, the landscape dried out, including Lake Victoria. With the loss of the lake, most of the descendants of the haplochromine cichlid hybrids presumably perished. Meier told me that she thinks a few hardy species managed to persist, however, perhaps in isolated swamps that remained as the lake went through this and subsequent drying and filling cycles.

Starting about 15,000 years ago, as the world was warming after the last ice age, the rains returned to East Africa. As Victoria's dry lakebed began to refill, the descendants of the hybrid haplochromine cichlids were able to leave their refugia and recolonize the emerging lake. In their new home, they discovered a cornucopia of possibilities for places to live and lifestyles to pursue. Over the years, the surface of Lake Victoria grew larger than the surface area of Switzerland, and the lake included a huge range of habitats, including shallow and sandy coasts, rocky outcrops, and open water.

The colonizing haplochromine cichlids took extraordinary advantage of the largess of the lake, quickly developing into new species that occupied nearly every available habitat and lifestyle. Some evolved into zooplanktivores, which consumed tiny animals drifting in the lake's waters. Other species consumed small prawns that lived near the lake bottom. Some cichlids fed on algae they scraped off rocks, while others ate detritus (decaying organic matter) on the lake bottom. Still others lived off snails, with some developing shell-crushing teeth and others without crushing teeth removing snails from their shells. Some cichlid species grew larger and developed gaping mouths for swallowing other haplochromines whole.

As more new species were occupying more and more lifestyles, some haplochromine cichlids developed unusual specialties. One species, for example, specialized in eating the scales of other cichlids. These scale-eaters would sneak up behind an unsuspecting fish, scrape scales off the victim's tail, and swallow the protein-rich meal as they swam away. Other species became cleaners: they provided a service to other fish by removing their ectoparasites, getting a free meal in the process.

Perhaps the most extraordinary example of adaptation occurred with the *paedophages* (child eaters), which terrorized female haplochromines by devouring their broods. All of Lake Victoria's haplochromine cichlids are maternal mouthbrooders. When spawning, after a female cichlid releases her eggs into the water, nearby males quickly fertilize them by releasing sperm. As soon as the eggs are fertilized, the female sucks the eggs into her mouth, thereby shielding them from the many hungry predators lurking about. When her eggs hatch, the mother fish will allow the young fry to leave her mouth to look for food, but at the first sign of danger, she sucks them back in to relative safety. Paedophages specialize in outwitting the mother's protective instincts. One way a paedophage does this is by swimming up to a mouthbrooding female, covering her snout with its huge, rubbery mouth, and sucking the eggs and fry out like boba through a bubble tea straw.

When scientists started to look closely at Lake Victoria's haplochromines in the 1970s, they found a spectacular diversity of species. In addition to discovering many unique lifestyles, they discovered that each bay, each island, and each stretch of coastline seemed to have its own unique local species of cichlids. As the scientists cataloged more and more species, they published estimated tallies. An early report in 1974 estimated that at least 150 unique cichlid species occupied the lake. A few years later, other researchers estimated that at least 200 species were present. By the early 1990s, the estimate grew to 300, or even 400, species. The one-upmanship in estimates continued, with other researchers noting that at least 500 species must have evolved in the 15,000 years since the lake refilled. Scientists called this exceptionally diverse group of cichlids a *species flock*, defined as a group of closely related—and often difficult to differentiate—species that live and interact in the same area.

The biologists studying Lake Victoria in the 1970s and 1980s, including Goldschmidt in his leaky boat, were enthralled and perplexed. How many species of fish were swimming in these waters? How could so many evolve in only 15,000 years? And what can be learned about evolution more broadly from these fast-evolving fish? They scrambled to study the lake in an effort to unlock the secrets of these remarkable Lake Victoria cichlids.

EXUBERANT EVOLUTION

Several generations of biologists have tried to answer the question of why so many species of haplochromine cichlids arose in Lake Victoria in such a short period of time. Although not all researchers agree on the exact recipe, they usually emphasize some combination of four attributes: the haplochromines were ecologically versatile, the cichlids that colonized the lake had unusually high genetic diversity, adult cichlids didn't stray far from their parents, and females were choosy about the appearance of their breeding partners, which responded by altering their colors and patterns. These factors were aided by the fact that the cichlids serendipitously colonized a very large and complex lake.

The haplochromines that swam their way to Lake Victoria when the waters returned were probably ecologically versatile generalists—jacks of all trades in the natural world that can eat a variety of foods and thrive in a range of habitats, according to one of the world's foremost experts on the evolution of Lake Victoria cichlids, Professor Ole Seehausen of the Bern University Institute of Ecology and Evolution and the Swiss Federal Institute for Aquatic Science and Technology. Seehausen believes that when these cichlids colonized the lake, they were able to survive in many of its diverse habitats, quickly finding food along the rocky shores and sandy bottom, even in deepest darkest parts of the lake.

For mouthbrooding haplochromines, the apple doesn't fall far from the tree—or, perhaps more accurately, the fry don't swim far from their mother's mouth. Unlike corals, which attempt to reduce predation by sending their offspring into open water so they can drift away, haplochromine young stay with their mothers until they are ready to fend for themselves. And once they're ready, they don't need to swim far away to find sustenance. By staying close to their mothers, haplochromines remain in neighborhoods of closely related cichlids that don't usually mix with cichlids from distant parts of the lake. Because Lake Victoria is so large, with many bays and islands, local populations of haplochromine cichlids quickly became reproductively isolated from one another. The lack of interbreeding with other populations enabled local populations to explore their own evolutionary paths, which may have differed from the paths of other populations.

Haplochromine females tend to be very particular about the appearance of their would-be suitors. In response, many males have evolved to become flashy and showy, especially around breeding time. In clear water, females can carefully evaluate their suitors for the best colors and patterns to get them in the mood. Females also tend to be somewhat fickle in their preferences for colors and patterns over time, and the males have responded by continuously offering slight variations on a theme for the females to evaluate. Consequently, male color patterns tend to change quickly from generation to generation, as female preferences drive the evolution of their colors and patterns into new directions.

In Lake Victoria haplochromines, female choice can even drive speciation within groups of species that are living in the same part of the lake. For example, females of closely related *Pundamilia* haplochromine species that are living on the same rocky reef have different preferences in breeding partners: shallow-water females prefer blue males that are less prone to being seen and eaten by birds, while deeper water females prefer red males that can flaunt their color without as much risk of aerial attack. Over time, two species have evolved side-by-side in these environments—a shallow-water species with blue males and a deeper water species with red males. The same kinds of evolutionary patterns may have developed elsewhere in the lake, with pairs of blue and red cichlids evolving repeatedly on isolated rocky reefs.

Local specialization and evolution in different environments are common in nature. In many cases, this has given rise to groups of closely related species, such as Darwin's finches in the Galapagos or the long-wing butterflies of the tropical Americas. But what has set Lake Victoria's haplochromine cichlids apart from most other species is the speed with which this has occurred. In just 15,000 years, Lake Victoria's haplochromine cichlids evolved from a few colonizing species, to hundreds of unique species. One scientist described this as "the most exuberant" example of this kind of evolution in the world.

Although these factors help explain how new species could emerge in Lake Victoria, they don't fully address the question of how so many species arose so quickly. Cambridge University's Meier, however, has discovered what she believes is the last piece of the puzzle: the original hybridization of the Congolese and Upper Nile cichlid species 150,000 years ago. When this hybrid was formed, it was endowed with essentially two complete sets of genetic variation (the amount of variety that exists within a population's genetic makeup)—one from each parent species.

Imagine, for example, that the gene pool is similar to the pieces in a LEGO airplane kit. You can follow the instructions to build the same airplane, over and over again. Evolution of existing genetic variation would be like using those same LEGO pieces to build a few slightly different variations of airplanes: one with longer wings, one with a shorter fuselage, another with a larger propeller, and so on.

Meier explained that when the ancestors of the Lake Victoria haplochromine cichlids hybridized, they essentially mixed two different sets of generic material. So, for example, imagine combining the kit for the airplane with a kit for a sailboat to create crosses between the two, such as seaplanes or jet boats. You could also use the LEGO pieces to make something entirely new, such as a motorcycle, using the wheels and engine from the plane and the hull of the boat. The hybrid ancestors of Lake Victoria's haplochromine cichlids had an abundance of genetic pieces to create a variety of species featuring different shapes, colors, sizes, forms, and functions. When these cichlids made their way into the refilled Lake Victoria 15,000 years ago, their unusually high genetic variance was reworked repeatedly under different conditions in different parts of lake, perhaps fueling the phenomenally fast evolution of new forms.

A SERIES OF UNFORTUNATE EVENTS

In the 1980s, in the midst of the scientific efforts to understand how the Lake Victoria cichlid diversity came to be, species started to disappear and fisheries for the haplochromines started to collapse. It was as though someone was firing a shotgun into the species flock, leaving only a very few survivors. What was causing the downfall of the haplochromines? Those who study Lake Victoria have a short version and a long version of the answer. The short version is that a giant predatory fish, the Nile perch, was deliberately introduced to the lake, and it gobbled up the haplochromines so fast that it overwhelmed the rescue effect for hundreds of species. The long version adds that, at the same time that the Nile perch were ascending, a host of other changes were also underway.

As its name suggests, the Nile perch is native to the Nile River in northern Africa; it's also found in many other freshwater rivers and lakes in northern and western Africa. It resembles a bass or a snapper—but super sized. The largest individuals can grow to almost 6 feet (2 m) long and can weigh as much as 440 pounds (200 kg), which is about the size of an adult male African lion. The Nile perch is also notable for its huge, suction-powered jaws, which help it swallow fish whole.

Although Lake Victoria is one of the sources of the Nile River, Nile perch were historically confined to the lower parts of the river by barriers such as waterfalls that kept them from swimming upstream to the lake. However, in the first half of the twentieth century, existing fisheries in Lake Victoria were deemed by British colonial leaders to be underperforming economically. They debated "whether fishery productivity might be increased by introducing non-indigenous species to Lake Victoria." Most of the fish being harvested were small and not suitable for markets outside of local communities. The idea was that, by introducing a large species of fish to the lake, the small fish—including the haplochromine cichlids—could support something more valuable: a harvestable fish with desirable flaky, white filets that could be sold commercially. Nile perch was an obvious choice for introduction, because it was already present in other parts of Africa and it could grow to enormous sizes. For those promoting the introduction of Nile perch, the possible negative effects on haplochromine cichlids were of little to no concern. To them, the haplochromines were small "'trash fish' of very little value."

Not everyone saw the proposed introduction of Nile perch as a good idea, however. Some fisheries biologists who were studying the lake worried that the species introduction could have unintended consequences and might not achieve the purported benefits. For example, in 1960, Geoffrey Fryer of the East African Fisheries Research Organization in Uganda argued that the rationale for introducing Nile perch to increase lake productivity was "based on ignorance of several fundamental biological concepts." The most notable concern was that the Nile perch would consume too many of the small fish, which were a potential food source for local communities. Fryer warned that "such an introduction is not only undesirable, but would jeopardize the existing commercial fishery."

As Fryer was warning about the risks of introducing Nile perch to Lake Victoria, it appears that the fish had already been secretly introduced to the lake. One author has since concluded that Nile perch were probably first introduced in the lake in 1954 by staff of the Uganda Game and Fisheries Department, though this wasn't documented until almost a quarter century later. By the early 1960s, the governments of Uganda and Kenya were publicly on board with the idea of officially introducing

Nile perch to the lake and conducted a series of deliberate releases, totaling a few hundred individual fish.

Many species introductions are not successful, as species can struggle
to adjust to life in an entirely new habitat. But the Nile perch seemed to
adjust easily to life in Lake Victoria. As early as 1962, some reports indicated that they were already breeding. Once established, the Nile perch
encountered nearly ideal conditions for rapid growth: the fish could
grow bigger than any other fish in the lake, outgrowing any potential
predators (other than people); the fish had access to an abundance of
food, including the haplochromine cichlids; and the fish experienced
no overcrowding with other Nile perch. As a result, the Nile perch population exploded. A survey conducted in 1969–1971 estimated that the
number of Nile perch had increased 10,000-fold since its introduction
to Lake Victoria. Although this was a dramatic increase, a lake of this
size takes a long time to colonize fully. Even after a decade of rapid population growth, the Nile perch was still a relatively small player in the
lake's ecosystem.

By the mid 1970s, however, significant numbers of Nile perch were
recorded in Ugandan fisheries, and Kenyan fisheries reported similar
increases by around 1978. For a few years, it appeared that the haplochromine cichlids might be able to hold their own, with continued
strong fisheries despite the rising numbers of predatory Nile perch. But
by the 1980s, it was like someone flipped a switch, and the number of
haplochromines tanked while Nile perch populations skyrocketed. One
scientific survey in Uganda noted that, in 1983, 76 percent of the lake's
catch comprised haplochromines; by 1985, haplochromine numbers had
declined substantially, to just 7 percent of the catch. Meanwhile, Nile
perch made up 16 percent of the catch in 1983 and 90 percent in 1985.
Apparently, Nile perch were quickly eating through the hundreds of
species of haplochromine cichlids.

Scientists immediately recognized that these rapid declines could put the future of the haplochromine cichlids at risk. By 1985, one group of researchers was already arguing that, "The establishment of [Nile perch in Lake Victoria] is not only an economic and ecological tragedy but an enormous loss to evolutionary biology." Scientists working in the Mwanza Gulf in Tanzania noted that, of the 123 species of haplochromine cichlids they had recorded in the area, fully 80 had disappeared by 1986—a 65 percent decline.

It's easy to look at the decline of haplochromine cichlids and see the Nile perch as the smoking gun; however, as the numbers of Nile perch were increasing, myriad other changes were underway in the lake. Intensive fisheries for the haplochromines and other species had probably already reduced their numbers in some places by the time the Nile perch populations began to increase. The one-two punch of Nile perch predation and human fisheries was probably more than some cichlid species could handle, and their numbers were driven so low that reproductive rescue would no longer be sufficient to help them recover. At the same time, increases in human populations in the Lake Victoria watershed were generating more pollution, which made the lake murkier and may have affected the availability of oxygen in the lake. To top it off, the Nile perch wasn't the only new fish introduced to the lake: several species of large, and in some cases predatory, fish had also been introduced to provide new opportunities for fishing.

In the past few decades, some scientists have concluded that, in addition to the rise of the Nile perch, changes in water clarity may have significantly affected the losses of distinct species in Lake Victoria. Matt McGee of Australia's Monash University told me that the murky water has "caused species to go extinct because they can no longer tell who to mate with." As with the *Pundamilia* species, females of different species in the same part of the lake may prefer males of different colors. One scientific study led by Seehausen concluded that, when water pollution "turns off the lights in the lake," females can no longer discern the colors of male suitors, leading to unintentional hybridization. According to McGee, the result may be that "what used to be two or three distinct species living in the same habitat with different color patterns is now kind of one thing." As a result, species diversity may decline, even if the genes that gave rise to those different species live on in the new hybrids.

From the perspective of the hundreds of species of haplochromines in Lake Victoria, pretty much everything in their world was changing. They had to cope with changes to their food resources, predation from introduced fish, increases in fisheries, and changes in conditions, such as pollution, that affected the clarity and chemistry of the lake. Faced with so much change at once, the haplochromine population's high-speed freefall in Lake Victoria was not surprising.

By 1992, Goldschmidt and other researchers estimated that 200 haplochromine species in Lake Victoria had either disappeared or were threatened with extinction. That's roughly the same number of marsupial species—including koalas, kangaroos, and wombats—in Australia. Professor Les Kaufman of Boston University, who started studying Lake Victoria haplochromines in 1989, described this as "the first mass extinction of vertebrates that scientists have ever had the opportunity to observe." In 1996, Goldschmidt wrote, "Never before in recorded history had the mass extinction of vertebrate organisms occurred on such a scale or within such a time frame." Scientists may never know exactly how many species were lost, because they hadn't come close to completing the monumental task of identifying and cataloging all of them by the time they began declining.

The introduction of the Nile perch also had enormous economic consequences. Dr. Richard Ogutu-Ohwayo, who has studied Lake Victoria since the 1970s and is currently coordinator of the African Lakes Network, told me that the rise of the Nile perch was accompanied by a tenfold increase in fish catches in Lake Victoria. The Nile perch were not only abundant, but they were also valuable, because they could be sold as part of an international export market that was worth about $250 million annually by the 2000s.

Of course, the huge changes to the fishery and composition of the lake created both winners and losers, economically. Ogutu-Ohwayo told me that much of the profit from the Nile perch fishery went to foreign investors, who built processing plants and exported the fish. But local fishers also saw new opportunities to tap into this market and its foreign cash.

Horace Owiti Onyango, a research scientist at the Kenya Marine and Fisheries Research Institute and a PhD student at Cornell University, has a personal connection to the Lake Victoria fisheries in Kenya. He grew up in a village on the Kenyan shores of Lake Victoria, and haplochromines were an important part of his diet. He recalls that, in the 1990s, when the Nile perch populations started to rise and the haplochromine populations were in decline, people in his village "felt quite threatened," because their staple foods were disappearing; worse yet, they were not interested in eating Nile perch, which they initially found unpalatable. This view largely changed, however, when people realized that there was money to be made catching and selling Nile perch.

Today, Owiti Onyango has noticed a generational divide with regard to the community's opinion of the Nile perch introduction. The older generation, exemplified by his grandmother, "lament" that they "can no longer access the species that they used to have." In contrast, younger people think that "the Nile perch is the best thing that happened to the lake," because of its economic value. "As a social scientist, I . . . identify with the earlier generation" and their concerns, he said, but "as an economist, I find sense in" the Nile perch introduction "because of the economic returns to the region."

In recognition of this economic value, the fisheries that boomed for Nile perch in the 1990s have become so valuable that fishers have scrambled to cash in by catching as many of the fish as possible. As the demand for the fish increased, fishing pressure arguably became too high, resulting in overharvesting. When a fishery harvests too many fish, catches will decline in the future, even if a reproductive rescue baby boom occurs. If fishers simply reduced their harvest rates, they might be able to make more money in the long run. However, it's difficult to get everyone to agree to relax fishing effort, especially in a lake shared by three countries, and intensive harvesting has now become the norm for Nile perch in Lake Victoria. Although overharvesting is generally considered undesirable in a fishery, in the case of the Nile perch, it may have unintentionally become the saving grace for the remaining haplochromines, many of which were poised for a comeback.

ORANGEHEADS

Taken as a whole, the number and variety of haplochromine cichlid species in Lake Victoria is difficult to comprehend. Each species, however, has its own story as it has struggled to adapt to the changing lake conditions. We will never know the details regarding how many were affected, because no one documented what happened. Even the estimates of how many cichlid species have survived in Lake Victoria—and how many have been lost—are extrapolations. One expert I spoke with described them as "guestimates."

Fortunately, there was an important exception to the lack of data on the haplochromines: A research group led by scientists from the Netherlands collected detailed information about how some of the haplochromines responded in Lake Victoria's Mwanza Gulf in Tanzania.

Importantly, these researchers collected information before, during, and after the Nile perch–induced crash. From these unique data sets, it is possible to follow the fates of some individual haplochromine species more closely.

Consider, for example, the haplochromine species that Dutch researchers originally called the "orangehead." Researchers noted that orangeheads were usually about 2–3 inches (51–76 mm) long, they lived in water 10–70 feet (3–21 m) deep over the muddy lake bottom, and they fed on passing zooplankton. In surveys conducted in 1979, orangeheads were found to be one of the most abundant zooplanktivore species in the Mwanza Gulf. When researchers returned in 1987–88, however, after the surge in Nile perch populations, they discovered that orangeheads were so rare that they were almost undetectable. In one sampling area they found none; in another area they found that orangehead populations had been reduced by over 99.99 percent since the late 1970s. Not only had orangehead populations severely declined, but the number of other zooplanktivorous haplochromine species also crashed. As a whole, the species group had declined by about 99.5 percent.

Clearly, by the late 1980s, orangeheads were on the verge of disappearing entirely from the Mwanza Gulf. Then, surprisingly, in the 1990s, their numbers increased. In surveys conducted in 1997, the orangehead populations had become nearly 18,000 times more abundant than during their low point in 1987. And their numbers continued to increase, so that by 2006, they were more than 10 times more abundant than they had been in the late 1970s, prior to the surge of Nile perch.

What happened to cause such a dramatic recovery? The most obvious explanation is that overharvesting of Nile perch in the 1990s removed most of the largest fish that had been preying on the orangeheads. Researchers believe that the smaller and less abundant Nile perch that remained were not as efficient at catching the haplochromines, and the few remaining individuals of fish species such as the orangeheads rebounded through reproductive rescue.

Reproductive rescue probably wasn't the full story, however, because not only did the orangehead numbers increase, but the fish actually changed in several ways. For example, the orangeheads of the 2000s occupied more territory than the same species of the late 1970s by expanding into shallower water. These fish were also eating new foods; in the 1970s they ate mostly tiny zooplankton, but by the 1990s, they were eating larger fish, shrimp, and mollusks. The typical orangehead fish was also more streamlined in appearance, with a smaller head and a larger tail, a shape that perhaps boosted its chances of out-swimming Nile perch. These changes to orangeheads occurred very quickly, within only a decade, indicating a very rapid shift in the behavior and appearance of the species. Remarkably, however, as Nile perch predation presumably declined because of intensive fishing, the orangeheads' bodies began to revert back to their original shapes, as they existed prior to the arrival of Nile perch.

Today, the orangeheads appear to have quickly achieved something of a détente with the Nile perch: their populations have adjusted quickly enough, likely through a combination of phenotypic rescue (changes in physiology, appearance, and behavior) and evolutionary rescue (changes in their genetic code) to survive. Moreover, their populations have grown; even in the continued presence of Nile perch, there are more orangeheads in the Mwanza Gulf now than there were in the 1970s.

RAPID EVOLUTION

Like the ancestors of the haplochromine cichlids that arrived in Lake Victoria, cichlid species are continuing to adapt to new environmental conditions in the lake. But what will happen to them going forward?

The most basic prediction from evolutionary biology is that survival of the fittest will favor those cichlids that have the best combinations of genes for dealing with new and rapidly changing conditions. This prediction, however, is largely based on a single species' path to evolution, whereby a given species slowly accumulates genetic changes over time to increase rates of survival. Some haplochromines appear to be doing something else as well. Remember that many species have been hybridizing in the lake as a result of changes, such as murkier water that makes it difficult for females to identify males of their own species. Though this recent hybridization is related to the net loss of species, it may also be creating an opportunity for the cichlids. As Ole Seehausen described it, "hybridization will be the source of new genetic combinations" for evolution to act upon, as happened with the haplochromine's ancestor.

In conservation, hybridization between two species is often considered a negative, because it means two separate species, each with its own inherent value, are merging into something entirely new. If both species completely intermix, with no survivors of the "pure" parent lineages, the net result is biodiversity loss: two species are gone, and one hybrid is created. Moreover, the hybrid itself is often perceived as being of relatively low value, because it wasn't shaped by natural selection over millions of years; instead, it's considered a mutt that resulted from an unnatural tryst.

But evolution doesn't view nature through this kind of normative lens; it's interested only in what works and what doesn't. What if the hybrid is actually better adapted to new environmental conditions? This may be what's happening today with the orangeheads in the Mwanza Gulf: evidence indicates that the fish are hybridizing with another species of cichlid. Given that these orangehead hybrids are thriving in the Mwanza Gulf, early evidence suggests that their hybridization may be giving them an advantage, or at least not creating a disadvantage, in the turbulent lake.

The orangeheads aren't the only cichlids thriving after hybridization. Jacco van Rijssel, of Wageningen University and Research Centre, showed me genetic data that he has collected on some of the resurgent species of haplochromines in the Mwanza Gulf. One species that specializes in eating mollusks in shallow water appears to be mixing with another species that tends to eat detritus, leading to at least two new populations with different ratios of DNA from the parent species. Some of the new populations prefer to eat predominantly mollusks with a little bit of the detritus, while others prefer the opposite. It's not clear whether these hybrids are new species that are arising as blends of the old ones, or the same two older species that now include a smattering of new genes from other species. In reality, there's probably no way to determine this exactly; instead, it's more of a philosophical question about where one species ends and another begins. What's clear, however, is that with all the change that has dominated Lake Victoria in recent decades, the haplochromines are already doing what they do best: building on their hybrid roots to adapt to new conditions.

In the midst of the haplochromines species' loss and hybridization, Seehausen said that one thing has been lost: ecological specialization. In stable ecosystems, species sometimes become highly specialized. For example, a species that focuses on certain kinds of prey can evolve specialized mouthparts that improve their ability to eat this prey. Specialization can lead to greater species diversity over time, as species branch out into separate ecological roles. Conversely, when the world starts to change quickly, being a specialist may prove to be detrimental. For a feeding specialist, this could mean that its preferred food is no longer available, so its specialized mouthparts are a disadvantage as it tries to find and eat different kinds of prey. This is one of the reasons why narrow specialists, rather than generalists, have become some of the world's most endangered species. For example, specialists such as polar bears, which live only in the Arctic and eat mostly seals, and pandas, which are confined to forests rich in bamboo that provides their only food, are much more vulnerable to environmental change than generalist black bears or brown bears, which can live in a huge range of habitats and will eat almost anything, including roots, berries, moths, and fish.

According to Seehausen, some of the surviving cichlids in Lake Victoria lack some of the feeding specializations of their recent ancestors. Perhaps this pattern is the result of rampant hybridization, in which relatively specialized parent species give rise to hybrid offspring that borrow some traits from each parent to create something in the middle: a generalist that can survive in a wide range of conditions on a wide range of prey. Instead of being an accident or an evolutionary dead end, hybridization within the haplochromines may turn out to be advantageous for dealing with turbulent changes.

The resurging species of haplochromines have achieved something remarkable: by adapting to changing conditions, they have largely returned to the same kinds of roles and the same abundance they experienced prior to the rise of the Nile perch. And this is also apparent in detailed data from the Mwanza Gulf, which shows that, in addition to the zooplanktivores such as orangeheads, mollusk-eating and detritivore populations were thriving by the mid-2000s. Moreover, representatives of most of the key feeding groups—such as algae grazers, insect eaters, and paedophages—had returned. Matt McGee of Monash University summed it up well, saying that, for many of the haplochromines, "life before and after the [Nile] perch is not that different. The only thing that's changed is [which predators are] hunting them down." Although many individual species have been lost to a combination of extinction and hybridization, the ecological community has been largely able to absorb these losses, as the species that remained are evolving into something new. By the early 2000s, though far less diverse by species, the commercial catches of haplochromines actually increased beyond the catches of the 1970s, and they have remained at this higher level, despite the fact that the Nile perch are now a permanent—albeit heavily fished—fixture in the lake.

SPECIES VALUE

Many people would argue that we have a moral responsibility to save species that are threatened with extinction. However, in practical terms, people decide whether and how to save a species based in part on their perception of the species' value. Although this is, by definition, a speciesist attitude—whereby people value one species over another—it's also consistent with the views of many toward wildlife in general. Consider this: If you only had enough resources to save one species, and you were presented with the option of saving either chimpanzees or one of the hundreds of unnamed and undescribed species of haplochromines, which would you choose? I'd wager that the vast majority of you would choose to save the chimps.

People assign value to species based on many different criteria, which often are related to some sort of personal bond or an opportunity for financial gain. The lost haplochromines arguably don't have much of either. If the haplochromines are known at all, only a small population of scientists and local fishers are familiar with them, which creates minimal opportunities to build a constituency. In terms of their economic value, these fish are primarily consumed and valued locally, rather than being part of a high-value export market. Owiti Onyango of the Kenya Marine and Fisheries Research Institute told me that people in his village who relied on the haplochromines as an important source of food did try to convince the government to do something to protect their historical fisheries. But in a lake the size of Victoria, what could fisheries managers realistically do? There would be no practical way to catch and remove every last Nile perch in the lake, even if they had wanted to do so.

The evolutionary biologists that I spoke with during my research emphasized the "existence value" of the haplochromine species, similar to the value assigned to mountain pygmy-possums. In other words, because the fish existed, they had intrinsic value as part of the world's biodiversity. They also noted the value of the haplochromines in Lake Victoria for studying evolutionary biology, because haplochromines are one of the most diverse and fast-evolving species flocks on Earth. However, these values weren't enough to fuel large efforts to try to save them once their populations started to decline.

Very little was done to save the haplochromines from extinction, according to Les Kaufman of Boston University. Instead, people largely left the haplochromines to their fate as they pursued the new fishery for Nile perch. Kaufman noted, however, that a few species were removed from the lake for captive breeding before they could go extinct. In the 1990s, he oversaw one of these efforts, which aimed to preserve thirty species of Lake Victoria haplochromines in zoos and aquaria. Although the project was successful for a while, the cost of maintaining the program was eventually deemed unsustainable, and today, only six haplochromine species are still being bred in captivity. Kaufman told me that the prospects of releasing these fish in Lake Victoria are poor, because somewhere along the way, they have contracted two new mycobacterial infections. These infections don't appear to originate from the lake, he said, so returning the fish would risk a lake-wide epidemic. He and his colleagues are now experimenting with housing the haplochromines in ponds in Florida. At least one other aquarium is maintaining a few species, and an informal network of hobby aquarists raise some species for their home aquaria. All of these efforts are small, however, relative to the scale of species loss, and none of these groups appears to be moving toward reintroducing the lost species into Lake Victoria.

With the return of some haplochromine populations to Lake Victoria in the 2000s, opportunities to catch the fish have increased, even with fewer available species. It appears that recovering some of the economic value of haplochromine fisheries did not require saving every species. Moreover, because Lake Victoria's cichlids evolve so quickly, today there are already signs that new kinds of haplochromines are arising. Based on their past rates of speciation, we shouldn't be surprised to see new species emerge in the decades and centuries to come, as fish evolve to fill newly created, and recently vacated, ecological niches. The next generation of haplochromine cichlids will not be of the same species that were present before the Nile perch introduction, but that doesn't make them inherently any more or less intrinsically valuable.

The story of the rise of the Nile perch and the decline of so many haplochromine cichlid species, Kaufman wrote, "has become a kind of ballad, warning of the dangers of meddling with nature." From this point of view, the story of Lake Victoria is a tragedy, with hundreds of unique species now extinct because of human meddling. In the past few decades, however, it has become clear that this story is also about the persistence and renewal of nature. When I recently asked Kaufman about his view in the 1990s that the changes to Lake Victoria amounted to a "catastrophe," he said, "My way of thinking about this has changed." He continued, "Every species is a beautiful and unique entity," but they're also ephemeral. "We also don't need to constantly shed tears over every entity that we can't find anymore" so long as the underlying processes that give rise to new forms of life are protected. Based on this kind of process valuation, the Lake Victoria story is not as much about the dangers of meddling, but about the awesome power of the rescue effect, which helps organisms successfully adapt to a changing world.

THE NEXT FLOCK IS TAKING FLIGHT

"What happened in Victoria is the kind of bellwether for what's going to be happening other places," said Matt McGee. The kinds of changes experienced in Lake Victoria are common to ecosystems all over the world, including the arrival of new species and changes to the physical environment. Although the haplochromines of Lake Victoria have an exceptional ability to adapt quickly, no species is ever frozen in time; each continues to adjust to the changing world around it. And the Lake Victoria cichlids are hardly alone in their ability to adapt quickly. Scientific literature includes a growing list of examples of species that are quickly undergoing phenotypic and evolutionary rescue in response to new conditions. For instance, a 2001 study noted that North American Chinook salmon populations that had been introduced to New Zealand were evolving to adapt to local conditions. Another study in 2006 showed that black snakes in Australia have rapidly evolved to deal with an invasion of toxic cane toads. Still other studies indicate that new plant species, some of which are facilitated by hybridization, are arising in Europe so quickly that the species creation rate is exceeding the extinction rate.

The response of the haplochromines since the 1990s demonstrates how strong the rescue effect can be in nature. In the face of a rapidly changing environment—new predators, pollution, and shifting food webs—many species have found a way to persist through a combination of reproductive, phenotypic, and evolutionary rescue. Furthermore, the genes of many extinct species live on in new hybrids that may provide the evolutionary seeds of new species yet to come. New ecological communities have also formed, with species such as orangeheads living in new habitats and seeking new prey.

Despite all these changes, the Lake Victoria ecosystem hasn't collapsed; it has simply shifted. Environmental change results in winners and losers, as some species wax and others wane. Because it contains so many species of fish, Lake Victoria has been able to reorganize quite effectively over a relatively short period of time, all the while providing enormous human benefits through fisheries. In our rapidly changing world, we can expect many other ecosystems to respond similarly: they won't stop functioning, but they will function differently. Extinction may well occur along the way, but the net result will be new ecosystems based on new conditions and groups of species.

Over time, nature tends toward diversification as the world accumulates more diversity of species. At a global scale, the world is unambiguously experiencing a period in which this trend is reversing—that is, a global extinction event with a net loss in the diversity of species. This does not necessarily mean, however, that we are reversing the underlying processes that generate new species. On the contrary; in many cases, we are accelerating these processes, because when species confront new conditions, perhaps because of climate changes or new species combinations, they will face new pressures to adapt. As a reaction to these pressures, a species may adapt by creating new forms and functions, like the orangeheads in the Mwanza Gulf. Or perhaps species or populations that have been long separated will intermix, creating a new generation of hybrids that may have traits that serve them well in the future, like the ancestors of the Lake Victoria haplochromines or the newly hybridized populations that have recently arisen. Or species will relocate to entirely new areas, where they may evolve apart from their ancestors. One such example of this involves a relative of Lake Victoria's haplochromine cichlids, *Astatotilapia bloyeti*, which was introduced to Tanzania's Lake Chala in the 1960s. After only fifty years in the lake, the species has formed at least nine different groups that may well be on their ways to becoming new species.

In conservation, the future may seem bleak if it is measured against a static historical standard. After all, Lake Victoria will never again have the mix of cichlid species that were present prior to the Nile perch introduction. But by a standard of its adaptive potential, Lake Victoria is cause for considerable optimism. The rescue effect is so strong that the species in the lake have already adapted and reorganized, and the future promises more adaptation to come as the next species flock takes flight.

CHAPTER 7

Scenes from the Pleistocene

THE THREE SPELEOLOGISTS WALKED gingerly, trying not to disturb anything on the bone-strewn cave floor. They could tell that this cave chamber was large from the delayed return of their echoes and the fact that their lights could not illuminate the far walls. Together, they had explored scores of caves, but this one was larger than anything they had ever witnessed in this part of southeast France. Already, it was yielding spectacular results, with beautiful rock formations and a wealth of bones.

Earlier in the day, the group had entered the tiny cave opening at the base of a cliff along a river gorge. Experience told them that most small caves yielded very little to explore—but no one could know for sure what they would find without looking. Once inside, they came to a pile of rocks that blocked their path, which could mark the end of their exploration. However, they detected a slight breeze coming through the rubble—a potential sign of a larger opening on the other side. So they carefully moved the rocks, clearing a space that was just big enough for them to wriggle through, one by one.

To the explorers' delight, the cave widened into a huge chamber. But a dangerous 30-foot (9 m) drop from the entry passage to the chamber floor below stopped their progress. Eager to explore their new find, they returned to their car to retrieve a rope ladder, which they used to lower themselves to the cave floor. Inching forward, they discovered that the floor was covered with the bones of ancient cave bears.

Finding the bones of extinct cave bears, *Ursus spelaeus*, is not uncommon in caves in this part of the world. Prior to their extinction about 24,000 years ago, these bears used caves across Europe and into Asia as shelter during winter hibernation. Some researchers believe that whole family groups of cave bears once hibernated communally in caves. Sometimes they died during hibernation, after which their bodies would decompose, leaving their skeletons behind. An abundance of bones indicated that this cave was once a particularly popular hibernation site. Within the sea of bones, the explorers noticed large depressions in the earth, which made the cave floor look as though it had been "bombed." They concluded that the depressions must have been "nests" that were carved out by the cave bears as crude beds for hibernation.

They passed the cave bear boneyard and traveled deeper into the cave. As they shined their lights on the surrounding cave walls, one of the explorers let out a cry of surprise. Something on the cave wall caught her eye—two unmistakable lines of red ochre. "*Ils sont venus!*" (They came!), she exclaimed. The speleologists knew then that they were not the first people to visit this cave. As they began to inspect the walls more closely, they spotted a drawing of a long-extinct woolly mammoth in red ochre.

Whoever had visited this cave and drawn this image had lived among these massive creatures.

Deeper in the cave, they found more drawings—some in ocher, others in charcoal—and etchings in the rock. One drawing featured a pride of cave lions, another showed a hyena standing next to a leopard, and another showed a herd of wild horses. Other drawings showed rhinoceroses locked in battle, more mammoths, and long-horned ibex. On one rock face, three giant cave bears were drawn, outlined in red ochre. They also identified drawings of reindeer, bison, giant deer, and aurochs, the wild predecessors of modern cattle.

The explorers found other signs of ancient human inhabitants, including charred remains of fires, some of which may have provided the light used by the cave artists. A child's footprints were preserved on the cave floor. Handprints and hand stencils adorned the walls. A charcoal drawing of a woman's pelvis and legs decorated a stalactite.

The people who inhabited this cave created a visual record of their world, emphasizing the grand animals that clearly loomed large in their daily lives and their psyches. Some of their subjects doubtless were sources of food, especially the smaller deer, aurochs, and ibex. But many of these animals would have been very dangerous for a human to encounter. Herbivorous woolly mammoths and rhinoceroses were enormous and could easily kill a human who was in the wrong place at the wrong time. Carnivorous cave lions, hyenas, and leopards would have killed and eaten people when they had a chance.

As the explorers were getting ready to leave, they paused for a moment to contemplate the significance of this discovery. They felt like intruders in someone else's space, later writing that "the artists' souls and spirits surrounded us. We thought we could feel their presence; we were disturbing them." The cave provided an intimate view into a different world in another time. Even looking at pictures of the cave drawings can feel voyeuristic, like prying open someone's private sketch book or photo album.

We now know that when the three speleologists—Eliette Brunel Deschamps, Jean-Marie Chauvet, and Christian Hillaire—entered what is now known as Chauvet Cave in December 1994, they were the first visitors—human or bear—in more than 20,000 years. Later explorations revealed that a rockslide had blocked a large entryway far from the crevice that the explorers cleared, sealing the cave off for millennia and preserving its artwork, artifacts, and bones until their chance discovery. Most of the artwork was created by people who lived in this area more than 30,000 years ago, making these among the oldest known drawings in the world.

PLEISTOCENE MEGAFAUNA EXTINCTIONS

The artists of Chauvet Cave lived during the middle of the last ice age, when the Earth was cold by modern standards. As a result, the ice sheets that had spread southward from the Arctic into North America and Eurasia were continuing to advance. The continent (today's Europe) was like a frigid version of the African Serengeti, with large beasts inhabiting an expansive grassland ecosystem, or steppe. The artists sketched these animals with extraordinary detail and accuracy, indicating direct personal experience. Because many of the species—including the woolly mammoths, cave lions, cave hyenas, woolly rhinoceroses, and cave bears—are extinct, these drawings are some of the only contemporary depictions of their appearance in life.

While Europe was losing some of these large animals, or megafauna, to extinction, similar extinctions were happening on other continents, including modern-day Asia, the Americas, and Australia, and on various islands. In fact, so many large animals disappeared near the end of the Pleistocene that scientists refer to the phenomenon as the *Pleistocene megafauna extinctions*. For centuries, scientists have puzzled over why so many large animal species went extinct during a relatively short period of time. They have proposed two main theories: the Pleistocene megafauna were victims of natural climate change as ice ages came and went, or these animals were driven to extinction worldwide by adept human hunters.

We know that the climate was changing, first cooling from about 110,000 years ago to the coldest part of the last ice age about 25,000 years ago, and then progressively warming until now, which makes it likely that climate change was a factor in many of the extinctions. As the climate changed, species would have needed to adapt rapidly. Many were probably undergoing phenotypic and evolutionary rescue to cope with changing temperatures, food availability, and predators. Many also would have been experiencing geographic rescue if they moved to new locations to find more suitable conditions. All this change could have made many species vulnerable to extinction as their population sizes declined while attempting to adjust.

The climate change theory alone is not a satisfying explanation for all of the Pleistocene megafauna extinctions, however. After all, these species had been through many rounds of oscillating cool glacial and warm interglacial weather, and most had persisted. Scientists have long argued that hominins—modern humans and closely related species such as Neanderthals—had something to do with the extinctions in Europe and Asia. And perhaps something similar happened as modern humans first reached the Americas and Australia. We know that these modern humans possessed intelligence and skills unmatched by any of their predators or prey: they could imagine complex attack strategies;

compensate for their relatively weak and tender bodies with tools made of wood, bone, and stone; and execute their plans with the aid of language to coordinate many hunters simultaneously. For species that were already reeling from a rapidly changing climate, the addition of hominin hunters may have simply overpowered the rescue effect.

The cave bear illustrates the scientific uncertainty about the causes of Pleistocene extinctions. Cave bears roamed throughout Europe and into Asia for hundreds of thousands of years. By examining their teeth and the chemical composition of their bones, scientists have determined that cave bears were mostly vegetarians. To maintain their enormous body weight (males grew to 1100 pounds, or 500 kg), they required a substantial plant-based diet.

Many well-preserved bones have been found in caves such as Chauvet, and these can be chemically analyzed to determine their approximate age. These ancient bones also contain DNA that can be decoded using modern biotechnology. By studying the bones of cave bears, scientists have determined that the animals' genetic diversity and population size were starting to decline about 50,000 years ago.

At this time, as the climate was cooling, glaciers would have been gradually sliding down mountains, covering what was once cave bear habitat. In some parts of Europe, the climate at lower elevations may have become too dry for the cave bear's preferred food plants to survive. One group of researchers suggests that the bears may have become trapped between glaciers in the mountains and the overly dry steppe grasslands below. At first, the bears seemed to hold their ground in between, on low mountain slopes. But their populations continued to dwindle, and by about 25,000 years ago (around the coldest part of the last ice age), most of them had disappeared. At least a few holdouts survived for another 1000 years in what is now Italy before eventually disappearing.

Although climate was undoubtedly an important cause of the decline of the cave bear, scientists have postulated that hominins were also involved. Perhaps the people who occupied caves like Chauvet deprived the bears of their much-needed winter shelter. Scientists have also discovered cut marks on cave bear bones, indicating that hominins at least occasionally butchered, and perhaps hunted, the animals. Similar to the tigers in Panna National Park, even a small amount of hunting of an isolated population of large mammals could have been enough to drive them locally extinct. As the cave bears tried to adjust to a changing climate, perhaps hunting by and competition with people was the final straw that drove them to extinction.

As with many of the Pleistocene megafauna species, no smoking gun has been found to explain their extinction. As scientists continue to gather more information and debate the merits of one explanation versus another, they have reached one consensus: the world's diversity of large land animals took a huge nosedive between the late Pleistocene and the current era. And without key predators and herbivores, many of the world's ecosystems changed. The surviving smaller predators and herbivores became dominant, vegetation shifted with different grazers and a changing climate, and ecosystems settled into forms more familiar to us today, such as forests, rather than steppe grasslands, throughout much of Europe.

RESURRECTIVE RESCUE

With so many members of the Pleistocene megafauna lost to extinction, the world is distinctly less rich biologically. Recently, a vocal group of conservation advocates began to consider whether some of this lost richness might somehow be resurrected. After all, traces of many species can still be found in DNA extracted from ancient bones or animals frozen in glaciers or permafrost. Instead of accepting these ancient extinctions as *fait accompli*, some advocates wonder whether some of the species, or even entire ecosystems, could be re-created. Accordingly, researchers are starting to experiment with what I call *resurrective rescue*, an entirely new approach to saving life on Earth. This human invention involves using the tools of biotechnology to try to bring extinct species back to life, à la Jurassic Park. Advocates of restoring lost species often use the term *de-extinction* to indicate that they are trying to reverse species extinctions. However, at best, the currently available methods could bring back only an approximation of the extinct animal. In other words, it's not a reversal of the extinction process, but rather an effort to create something new based on the lost species as a model.

The woolly mammoth is the most commonly cited candidate to resurrect from the Pleistocene extinctions. Mammoths were ancient elephants, but unlike today's Asian and African elephants, they lived in cold, northern latitudes. Their shaggy coats helped insulate them from the cold, and their ears were smaller than those of modern elephants, which helped them retain heat. At their peak, they ranged all across northern Europe, Asia, and North America. Like cave bears, mammoth populations declined in the waning days of the Pleistocene, with the last few individuals dying on Wrangel Island in the Arctic Ocean about 4000 years ago.

The idea of encountering long-extinct species such as mammoths or cave bears has long been relegated to fantasy books and movies. In the last few decades, however, as some of the requisite tools for reading and manipulating DNA have become less expensive and more effective, the science fiction of species resurrection is steadily marching toward being science, sans the fiction. Even so, advocates of resurrecting extinct species still have an enormous Humpty Dumpty problem in determining how to put the pieces back together. In the case of extinct animals, the pieces are fragments of ancient DNA, and the currently available tools that could be used to accomplish this are cloning and gene editing.

The clone of the cave bear

In genetics, an organism is considered a clone if it is genetically identical to another organism from which it was derived, even if the two individuals are not superficially identical. For example, the many thousands of polyps that live in a large coral colony are genetic clones, even though they vary in size and/or age.

Inside each animal cell, a nucleus stores most of the animal's DNA. Scientists have developed methods for removing the nucleus of a living cell and replacing it with the nucleus of another cell through a process called *nuclear transfer*. In 1997, scientists made headlines when they announced that they had succeeded in cloning a mammal, a sheep later named Dolly. To create Dolly, researchers started with a cell from a living ewe (Dolly's first mother) and an unfertilized egg from another ewe (Dolly's second mother). Scientists then performed a nuclear transfer by removing the nucleus from her second mother's egg cell and replacing it with the nucleus from her first mother's cell. This new cell then developed into an embryo that was artificially implanted into a third ewe's womb (Dolly's third mother). The embryo eventually developed into a fetus and was born alive. The lamb, young Dolly, shared the nuclear genetic code of her first mother, the donor ewe. In other words, Dolly was a clone of her first mother.

Theoretically, the same process could be used to create a clone of a long-extinct species, such as the cave bear. However, three key ingredients would be needed to accomplish this: a donor egg cell that is similar to that of a cave bear, a surrogate mother that is similar enough to provide a suitable development environment (you couldn't grow a large cave bear cub inside the womb of a mouse, for example), and an intact nucleus of a cave bear with its entire genetic code. Spoiler alert: the first two ingredients may be available, but the third is not.

Consider a hypothetical example about the possibility of cloning a cave bear. The cave bear was closely related to today's brown bears. Therefore, it is possible that scientists could work with that species to harvest donor eggs that could then be used to artificially impregnate surrogate moms.

Cave bears evolutionarily split from brown bears about 1.5 million years ago, which is not especially long by evolutionary standards. In fact, evidence from DNA extracted from cave bear bones indicates that they hybridized with brown bears as recently as a million years ago. The closer the donor and surrogate are to the species to be cloned, evolutionally speaking, the more likely the cloning procedure could be successful. If one assumes that a brown bear could provide a suitable egg and womb, the last ingredient required to create a clone of a cave bear is a cave bear cell nucleus. Unfortunately, such a nucleus does not exist: the bones of cave bears found in caves contain a lot of DNA, but the structures that held the nucleus and DNA together have long since broken apart.

Dolly-style cloning may be used in species conservation efforts, but only if a suitable nucleus is available. For species that are not extinct, living cells can be used as the source for the nucleus. For example, gray wolves have been successfully cloned this way, using living wolf cells and domestic dogs as egg donors and surrogate mothers. Similarly, researchers have tried to clone endangered Bactrian camels using common dromedary camels as egg donors and surrogate mothers, but these efforts have not yet yielded viable Bactrian camel clones.

This same approach might also be used for extinct species if some of their living cells are available. For example, researchers cryopreserved cells from the last living bucardo, a Pyrenean ibex, before she died. They have since attempted to clone the animal using nuclei from the cryo-preserved bucardo cells, eggs from domestic goats, and surrogates from various ibex and ibex/goat hybrids. As of this writing, only a single pregnancy has gone to term, but the young bucardo died immediately after birth. If ultimately successful, a cloned Pyrenean ibex could be the first successfully cloned extinct organism.

Engineering a precision hybrid

Without the availability of intact nuclei, can these ancient species ever be resurrected? Perhaps. Scientists can use imperfect ancient DNA as a biochemical recipe of sorts to re-create nuclei that share some genes with the extinct species.

Let's consider the cave bear example to explore how such a recipe might be created. Scientists have extracted a lot of cave bear DNA from bones found in European caves. The DNA is heavily degraded, how-ever: the extremely long molecules of DNA have broken down into mil-lions of short pieces. By decoding these pieces, scientists can identify a series of DNA phrases. To illustrate, suppose you cut up many copies of this book into snippets of several words. A computer could be used to re-create sentences by looking for places where the snippets overlap. If the computer found the segment, "suppose you cut up many copies of this book," and found another segment from a different book, "copies of this book into snippets of several words," it would note the overlapping words and deduce that these two fragments are probably part of the same sentence.

The process of looking for overlapping DNA fragments is similar in a way, but it's complicated by the fact that the ancient DNA is certain to be contaminated with the DNA from other organisms, such as the bacteria and fungus that decomposed the cave bear's body, or even the DNA of the people who touched the bones while collecting them. The snippets from the cave bear's DNA would be mixed with snippets of DNA from different organisms, making it difficult to distinguish the cave bear DNA from the other DNA. "It's like the world's worst jigsaw" puzzle, said ancient bear DNA expert Axel Barlow of Nottingham Trent University. Barlow has reconstructed the DNA puzzles of cave bears from bones collected across Europe and in parts of Asia.

Reconstructing the cave bear's genetic code could be simplified by comparing it to the DNA of closely related species, which can be very similar. For example, Barlow said the sections of DNA from ancient cave bears that he studies are more than 99 percent identical to the sections of DNA of modern brown bears. By studying the DNA of this closely related species, he can more easily determine whether each new DNA fragment belongs to the cave bear. With a lot of samples and a lot of effort, scientists could eventually develop a reasonably complete cave bear genetic code, filling in any missing pieces from the DNA of a brown bear.

With the genetic code of both bears in hand, researchers could compare them to determine their differences. Then they could use CRISPR gene editing to edit the genes in a living brown bear cell to match the genes of a cave bear. CRISPR, a new family of genetic engineering tools, enables researchers to locate a specific part of an organism's DNA, remove it, and then replace it with different DNA. CRISPR could theoretically be used to remove genes from a living brown bear nucleus and replace them with genes that code for a cave bear. I describe this as theoretical because of the scale of changes required. If only one or two genes need to be changed, cave bears, mammoths, or other extinct animal nuclei could possibly be re-created relatively soon. But there may be millions of genetic differences affecting hundreds or thousands of genes that would need to be altered to change a brown bear nucleus into that of a cave bear.

Nonprofit organization Revive & Restore has become one of the leading advocates for resurrecting extinct species. For example, the organization is currently working in a partnership with Harvard University Professor George Church to edit the nuclei of Asian elephant cells to resemble those of woolly mammoths. Bridget Baumgartner, an expert in molecular and synthetic biology and a program manager at Revive & Restore, told me that the differences between the Asian elephant and woolly mammoth genomes comes down to about 0.16 percent of their DNA—in other words, the two species' genomes are about 99.84 percent the same. Furthermore, only a subset of these differences was probably important in differentiating Asian elephants from mammoths.

Baumgartner said that the organization's partners have identified "50 candidate genes that they think would be important for the cold," such as genes for long fur or for the ability of blood to transport oxygen in extremely cold weather. By focusing on this subset of genes, researchers are aiming to create a "precision hybrid," she said, a new kind of organism that is mostly Asian elephant but with targeted genetic traits of woolly mammoth that would enable it to survive in extremely cold climates.

If researchers succeed in editing the targeted genes, Baumgartner said they would proceed iteratively, checking to determine whether their edits created the desired effects along the way. And they could conduct many of their tests using cell cultures, rather than entire organisms. For example, they could grow clumps of skin cells to see if they produce long, thick fur. But even if researchers are confident that they have correctly identified the targeted genes and the cells are functioning as expected, they still face formidable challenges. First, they'd need to move the gene-edited nuclei into an elephant egg cell. Then they'd have to stimulate the egg into developing. And, finally, they'd need to be successful in planting this embryo in a suitable womb to produce a viable mammoth-like elephant.

THE ETHICS OF RESURRECTION

Resurrecting an extinct species through cloning and gene editing—if it is ever achieved—will never be easy. When Dolly the sheep was cloned, the researchers didn't succeed the first time they tried: in fact, they used 277 eggs, 29 embryos, and 13 surrogate mothers to eventually create one successfully cloned sheep. To date, efforts to clone a bucardo have been similarly inefficient without yet producing a viable clone. Although these processes have been refined over the years, the best-case scenario for cloning still includes a lot of failures, as each species poses its own unique challenges.

For long-extinct species, the challenges of resurrective rescue will be much more significant, because researchers won't be starting with a perfect cell nucleus to insert into an egg from the same species. Organismal development in nature is miraculous. We know from human medicine that an extraordinary number of things need to go right genetically and environmentally to produce a healthy infant. Even small defects in the genetic code or imperfections in the environment in the womb can cause miscarriages, stillbirths, and a host of serious genetic and developmental disorders. These kinds of problems already plague cloning via nuclear transfer. Adding the nearly blind alchemy of selective gene editing will only make these kinds of negative outcomes more common.

Presumably, researchers would approach resurrecting a gene-edited species iteratively, whereby new combinations of edited genes would be tried sequentially in combination with a variety of genetic insertion and artificial impregnation methodologies—in other words, using trial and error. Some of this trial and error could happen in cell cultures, but eventually it would have to be attempted with complete organisms with real-world consequences. How many dead and deformed mammoth-like elephants would be acceptable to create in pursuit of the goal of resurrecting this species? Ten, a hundred, a thousand? Would those with disabilities that didn't die before birth be coaxed into surviving as long as possible to advance scientific study, or would they be immediately euthanized? Furthermore, is it ethical to involve living, breathing Asian elephants in this process, given that this species is having conservation challenges of its own from lost habitat, poaching, and disease?

When I asked Baumgartner whether Asian elephants should be used as surrogate mothers for this process, she responded, "I would say it would be a non-starter altogether if this is a species on the brink of . . . extinction." However, she noted, there may be another option: an artificial womb, in which an embryo could develop to term in a human-constructed environment. Baumgartner said an artificial womb would be "a great option here, because it doesn't involve any elephants." Though on the downside, she admitted, "That technology is . . . in its infancy." Nevertheless, she is holding out hope that by the time researchers are ready to attempt to produce an elephant/mammoth hybrid, Asian elephant populations will be stable or increasing so they can be used as surrogates "without causing undue harm," because "using a natural surrogate would make this a lot more feasible, technically."

Extensive efforts to resurrect a species would likely be quite expensive, which raises a reasonable question about whether exploring resurrective rescue is the best way to apply limited conservation funds. Would these dollars be better spent trying to save living species from going extinct in the first place? The process of resurrecting a species, either with a surrogate or an artificial womb, would also likely be messy and potentially cruel to animals who suffer in the process. It's no surprise, then, that some people oppose the methodology, even if they harbor curiosity about the possibility of restoring a species.

WHY RESURRECT A LOST SPECIES?

One of the most important consideration in species resurrection may be the question of *why*. Reasons for resurrection could include pure curiosity, righting a past wrong, and rehabilitating past ecosystems.

While obviously fictional, the Jurassic Park franchise illustrates the curiosity argument for resurrecting species. People are fascinated by the amazing beasts that once roamed the Earth, and at least in the context of the movies, they are willing to pay for a chance to view them. Imagine the response if a creature resembling a woolly mammoth were resurrected. People would scramble for a chance to get a glimpse of it, even if their viewing was confined to a zoo.

A species may also be resurrected to try to correct a perceived wrong, especially if humans are implicated in its extinction. Just as one could argue that it's immoral to cause the extinction of a species today, one could argue that past human-caused extinctions were immoral and should be corrected if possible. Following this logic, all extinctions in which human involvement was suspected—such as many of the Pleistocene megafauna extinctions—would be candidates for resurrection.

Finally, many advocates for resurrecting species argue that people should focus on resurrecting species that played an important ecological role, which, because of their extinction, is now vacant. Beth Shapiro, an investigator at the Howard Hughes Medical Institute and a professor of ecology and evolutionary biology at the University of

California, Santa Cruz, described this argument as follows: It is "eco-logical resurrection, and not species resurrection, that is the real value of de-extinction. We should think of de-extinction not in terms of which life form we will bring back, but what ecological interactions we would like to see restored."

This kind of argument is often applied to mammoths, which, if resurrected, might be able to help slow global warming by playing their normal role in the ecosystem. To understand how this could work, first imagine how ancient mammoths may have shaped the ecosystems in which they lived. Like modern elephants, they were voracious herbivores that devoured a wide range of grasses, shrubs, and trees, while trampling any plants in their path. Scientists know from modern ecosystems that intense grazing and disturbances like those created by large herbivores can create grasslands in areas that would otherwise become forests. Mammoths probably helped create and maintain their steppe ecosys-tem, which in turn supported the remarkable abundance and diversity of large animals captured by the artists in Chauvet Cave.

Mammoths, which didn't hibernate, had to find food throughout the winter, even if it was covered by deep snow. To get a meal in winter, mammoths used their giant feet and tusks to dig through the snow to uncover the plants below. Similar to the mountain pygmy-possums in their underground nests, when the insulating snow cover was disturbed, the ground underneath would get colder in the winter, freezing more deeply. All of this freezing created permafrost—permanently frozen soils that trap carbon in the form of dead organic matter.

When the mammoths and other large grazers disappeared, so did their grazing and snow trampling. Over time, the steppe disappeared too. In warmer places, such as most of Europe, much of the steppe was eventually replaced by forests. But in far northern places such as Siberia, where permafrost still forms, the steppe was replaced by a new kind of ecosystem that has less grass and more moss, shrubs, and trees. This new Siberian ecosystem doesn't freeze as much as the mammoth-dominated steppe because of the insulating effects of mosses and undisturbed snow. The shrubs and trees are also sometimes tall enough to rise above the snow. Since they are darker than the white and reflective snow, the shrubs absorb more sunlight, further warming the ecosystem.

Today, permafrost in the Arctic is melting as a result of a combination of global warming, undisturbed snow and moss cover, and heat-absorbing plants. When permafrost melts, it releases carbon dioxide and other greenhouse gases into the atmosphere, accelerating global warming. The permafrost in northern Siberia alone is estimated to contain more than twice the carbon that is currently held in all of the world's rainforests combined. Some people believe that by restoring mammoths and other large grazers, the steppe could return, along with its deeply frozen permafrost, slowing the release of carbon into the atmosphere and perhaps slowing climate change. They argue that mammoth-like creatures, if released in places like Siberia, could provide this service to humanity.

Russian scientist Sergey Zimov is so taken with the idea of re-creating the steppe that he established Pleistocene Park, an experimental nature reserve in Siberia. Within the park, Zimov and his colleagues are rebuilding an ecosystem that resembles the steppe grassland ecosystem and populating it with megafauna species that survived the Pleistocene. Some species were already present here, such as reindeer and moose, but Zimov is also deliberately importing and introducing other large grazers such as muskox, horses, and bison to the park. Mammoths might also be helpful in re-creating the steppe, however, because their great size and weight would enable them to eat shrubs and trees and disturb the snow more than the living species.

The idea of rebuilding an ecosystem's function to slow global warming is compelling. Earth's climate is changing so quickly that it is worth pursuing many climate mitigation strategies at once. But before we move forward with resurrecting mammoths for this reason, it's worth asking whether such a project is practically feasible. For example, how long would it take to create the first mammoth-like organism? Realistically, it would take decades, while we wait for more advances in biotechnology, and because of the slow reproductive process of elephants. And if the resurrected organisms survived, how long would it take to breed a giant herd of mammoths to help transform the arctic back into steppe? Baumgartner told me that "this is where we get to the true rate-limiting step."

Imagine that researchers were wildly successful and produced twenty healthy baby mammoth-like elephants within a decade. It would then take another decade for them to become sexually mature, at which time a large-scale breeding program could be launched to populate the Siberian tundra. The maximum population growth rate for African elephants is about 7 percent per year. If we generously assume that the first twenty animals were healthy enough to reproduce at this maximum rate, then within 100 years (20 years to make the first twenty adult mammoth-like elephants, and 80 more years of exponential population growth through breeding), there would be 4484 living mammoth-like elephants.

Zimov believes that about 1 million square kilometers (400,000 square miles) of northern Siberia steppe could be restored to help slow climate change. He and his colleagues have also estimated that steppe ecosystems once supported around one mammoth per square kilometer. Thus, to restore the full ecosystem function of mammoths in northern Siberia, as many as a million mammoths may then be necessary. Is it reasonable to make a climate argument for creating mammoth-like elephants if, even under the best-case scenario, only a small fraction (around 4 percent) of the mammoths needed could be successfully bred in 100 years?

Furthermore, could the existing large herbivores create the same environment? In other words, are mammoths really needed? Zimov told me, "I am sure it's possible to trample most of [the] snow in Siberia . . . without elephant[s]. Caribou, moose, bison, horses, muskoxen, snow sheep do this with pleasure and without salary." He said existing grazers can work especially well to restore the steppe environment in areas where the snow cover isn't too deep and grazers have access to water during cold weather. At Pleistocene Park, this is being accomplished by the current herds of animals.

Why not, then, just breed these existing species at a large scale across the arctic tundra? After all, many existing species could be bred for this purpose today. Zimov told me that the main problem is people. In particular, he cited three issues: In places with agriculture, people don't want their crops overrun by large grazers. People who breed domestic animals don't want them threatened by wild ones. (Zimov told me that reindeer herders often shoot wild reindeer on sight because they worry the wild animals will lure their domestic herds away.) And, finally, people don't want restrictions imposed on the number and kinds of wild animals that they hunt for meat. Perhaps rather than going through all of the effort to resurrect a mammoth-like elephant, it would make more sense to work on changing these social barriers to repopulating Siberia with existing large herbivores.

Unless you're playing a really long game, well beyond the coming century of intensifying climate change, it doesn't make much rational sense to resurrect mammoths as part of a portfolio of climate-change solutions. Psychologists have described a human phenomenon called *motivated reasoning*, which operates below the level of human consciousness. If people are motivated to achieve a particular outcome, such as resurrecting mammoths, they can construct reasonable-seeming justifications for pursuing this outcome. Perhaps some advocates are allowing their desire to resurrect mammoths to drive their climate-change rationale, rather than the other way around.

That said, if abundant mammoth-like elephants could eventually be bred and released, they would certainly impact the ecosystem, but by the time the population reached adequate numbers, it would be far too late to solve our current climate change crisis. Furthermore, the ecological function argument is not unique to mammoths and could be made for resurrecting almost any species. For example, herbivorous cave bears probably helped disperse the seeds of vegetation they consumed, which passed through their digestive tracts and were deposited in their stool. They also may have used their giant claws to dig up roots to eat. This soil disruption could have affected which plant species would grow in certain areas. Could we then arrive at the conclusion that the ecological role of cave bears is so important that they should be resurrected and released to the wild?

RESURRECTIVE RESCUE IS OUR CHOICE

Despite our best intentions, extinction works in only one direction: all species will eventually go extinct, but extinction can never be truly undone. Perhaps someday scientists will master resurrective rescue and create new species that approximate those that have been lost. If they are successful, their creations may fill long-vacant ecological roles, such as helping to graze and trample the tundra to return the area to steppe grasslands. However, just as people cannot ever truly de-extinct long-lost species, people cannot perfectly re-create ecosystems of the past; too much has already changed in our climate, and there's no going back. Beth Shapiro put it best when she noted that the mammoth is "an icon of an ice age, and we are not in an ice age." Even if similar groups of species are eventually re-created, there's no reason to believe that, under different circumstances, the ecosystem will function as it once did.

Ultimately, this means that the world documented on the walls of Chauvet Cave is irrevocably gone. The Chauvet artists did not write a recipe for an ecosystem of the future; they recorded scenes from an ecosystem that is now in the past. Change is nothing new in nature, because all ecosystems and species are ephemeral. Indeed, this is exactly why the rescue effect exists: because life persists only when it can navigate an ever-changing world.

It is probably a good thing that efforts to resurrect long-lost species remain in the early stages of manipulating single cells rather than entire organisms. This gives society an opportunity to ask the hard questions about whether the goal of resurrecting these species should even be pursued. Nature, after all, does not care about the Pleistocene megafauna extinctions—or any other extinction for that matter—and nature cannot bring an extinct species back to life. As the creators of resurrective rescue, we are the ones who must decide whether and how to wield our growing power to resurrect life on Earth.

CHAPTER 8

Ascending to New Heights

IN 1815, Rear-Admiral Sir George Cockburn was entrusted to transport a passenger of singular importance, Napoleon Bonaparte, who had been recently taken into custody by the British government for the second time. Less than a year earlier, a coalition army including British, Prussian, and Russian soldiers had successfully captured the city of Paris, forcing Napoleon to abdicate his rule. To prevent him from returning to power, the British had exiled Napoleon to Elba Island in the Mediterranean. However, the wily emperor remained on Elba for only ten months, escaping and returning to France in February 1815 to reclaim his empire.

Upon his return, Napoleon's support in France grew rapidly, and his army was once again positioned to be a major military force in Eurasia. Concerned about what that might portend, European military powers quickly reassembled their coalition and readied themselves to confront the emperor's forces. But Napoleon's second reign was to be short lived, as he suffered a crushing defeat at the hands of the British and Prussians in Waterloo in June 1815 and scampered back to Paris. Adding insult to injury, the French quickly soured on him after his defeat, and lacking support, he was forced to abdicate his rule yet again.

As they had done the previous year, the British had to decide what to do with the deposed emperor. Remarkably, they followed exactly the same playbook: Napoleon was to be exiled to a remote island. Apparently having learned at least some lesson after his escape from Elba, they chose a location far more remote: Saint Helena, a small volcanic island in the South Atlantic Ocean, about 1200 miles (1950 km) from the coast of southwestern Africa.

Cockburn's mission was to escort Napoleon to Saint Helena and set up whatever means were necessary to ensure that he did not escape again. Upon arrival, the admiral had established a garrison on the island that was charged with repelling any attempt to help Napoleon escape. Cockburn also decided that uninhabited Ascension Island, 800 miles (1300 km) to the northwest of Saint Helena, posed an unacceptable risk, because it could be used as a staging ground before or after an escape attempt. So he unilaterally decided to take possession of Ascension Island for England. In a letter to the British Admiralty announcing the safe delivery of Napoleon to Saint Helena, Cockburn wrote that the purpose for occupying Ascension was "to prevent America or any other nation from planting themselves there as upon a hitherto unoccupied and unowned island . . . for the purpose of favouring sooner or later the escape from hence of General Buonaparte."

Accordingly, Cockburn dispatched two ships from Saint Helena with a secret order: "You are hereby required and directed to make the best of your way to the Island of Ascension, and in the event of finding the said island (as is probable) unoccupied by people of any nation whatsoever, you are to take possession of it, hoisting upon it the English flag, and putting ashore upon it an officer and about ten men and a gun." He continued: "You will then take measures to ensure that every boat or vessel which may approach Ascension be minutely examined, and should there be discovered on board any such, either General Napoleon Buonaparte or any French persons who accompanied him to Saint Helena, he or they, as the case may be, must be immediately secured and taken."

Cockburn couldn't have known than his military decision to occupy Ascension Island would lead to a settlement that lasted long past the death of Napoleon on Saint Helena, and that indeed survives to the present day. The admiral's decision would also have far-reaching consequences for the island's flora, as the British slowly began what has become one of the most laborious, and arguably successful, terraforming projects on Earth.

A RUINOUS HEAP OF ROCKS

In geological terms, Ascension is a "young" island of about one million years old. Like Africa's Lake Victoria, it is situated within a rift zone, where the Earth's tectonic plates are pulling apart from each other. This rift zone runs the entire length of the North and South Atlantic Ocean. In fact, it is the reason that the Atlantic Ocean exists: the rift pushed the Americas to the west, away from what are now Europe and Africa in the east. In the process, the volcanic rift zone created the largest mountain range in the world in the middle of the Atlantic. However, with only a handful of exceptions, such as the Azores in the north and Saint Helena and Ascension in the south, these mountains are located completely below the ocean's surface.

At 8 degrees latitude south of the Earth's Equator, Ascension Island has a tropically hot climate, with average daily temperatures rising to 95°F (35°C). When seafarers visited Ascension prior to the mid-nineteenth century, they found a rocky and nearly vegetation-free landscape, and most were not favorably impressed. Imagine if Yelp had existed back in the heyday of ocean exploration. Would anyone have ever gone out of their way to visit a place with reviews like this?

"The place was of no use as far as we could tell, and we left it behind us."

"Like a land thatt [sic] God has cursed."

"It was a ruinous heap of rocks."

"One may as easy walk over broken glas [sic] bottles as over the stones, if the foot deceives you[,] you are sure to get cut or lamed."

"I never saw a more disagreeable place in all the world."

"In all directions, nothing but the most barren and desolate region met our view."

"Any body would have believed that the Devil himself had moved his quarters and was coming to keep Hell on Ascension."

Clearly, Ascension Island lacked the lush foliage that covered other tropical volcanic islands such as Tahiti and Hawai`i. But why was it so desolate? One reason for its lack of vegetation was its relatively young geological age. It takes a long time for lava rock to weather and produce soil in which plants can take root. Today, parts of Hawai`i that have been scorched by recent volcanic eruptions are similarly barren. In addition, although constant breezes and low-passing clouds brush the top of Ascension's highest peak, rain showers come infrequently and quickly evaporate or trickle into the craggy rock. This is especially true on the island's lowlands, where one scientist estimated the total annual rainfall at only about 5 inches (13 cm). Finally, Ascension Island is highly isolated: the closest part of Africa is about 1000 miles (1600 km) away, and the coast of Brazil is more than 1400 miles (2300 km) away. If Ascension were closer to other islands or land forms, chance events may have helped plants colonize there. But with no close neighbors to serve as sources of plant life, Ascension remained barren. When it was first sighted by a Portuguese ship captain in 1501, young, dry, and remote Ascension was home to only a handful of plant species.

Early scientific explorers consistently found only five types of plants—and no trees—across the entire 34 square miles (88 square km) of the island. The meager flora eked out a living on an island largely comprising piles of sharp volcanic rock. In 1752, a visitor who braved a hike up one of the ridges on the island noted, "Neither on the sides, nor on the top, did I meet with one single plant." However, plants did grow atop the very highest peak, including knee-high ferns that soaked up moisture from passing clouds. In fact, this verdant patch was so exceptional to Ascension's barren landscape that the peak was named Green Mountain, signifying its contrast with the rest of the island. Later explorations identified about twenty-five total plant species on the island, many of which were diminutive and easily overlooked, such as species of moss and liverwort growing in cracks and crevices in the volcanic rock.

On such a barren island, the crew that Cockburn ordered to occupy Ascension faced a hard existence. There were no trees to provide shade from the oppressive heat or wood for building shelters or cooking. The few native plants on the island provided little to no vegetable matter to eat, and with so little arable land, crops were difficult to grow. Most importantly, there was almost no fresh water. Wells dug in the lowlands were salty, and the men were able to find only a trickle of fresh water high up on the mountainsides, which must have fallen as rain or been collected as condensation. On the plus side, the island provided ample sources of fish, sea bird eggs, sea turtles, and a few hardy goats that had been introduced to the island centuries earlier by passing sailors.

The first settlement on Ascension was commanded by William Roberts, who immediately took steps to arrange shelter and created a defense plan. Cockburn had also ordered him to "select the most promising piece of ground . . . and cultivate it" to supply the colony with fresh vegetables. Roberts eventually located a piece of suitable land comprising about an acre, at an elevation of 2200 feet (670 m) on Green Mountain. In this small plot, the men planted carrots, turnips, and potatoes, thereby starting a decades-long effort to increase the diversity and productivity of plants on Ascension.

When Napoleon died in 1821, the original purpose of the British occupation of Ascension was nullified. According to historian Duff Hart-Davis, "By then, the island had taken on another role, that of a providing a sanatorium and a victualling base for ships engaged in suppressing the slave trade on the west coast of Africa." So the British stayed on, and throughout the next few decades, they continued to find reasons to maintain a population on the island. During this time period, a series of commanders steadily improved access to fresh water and cultivated more produce through hard labor and ingenuity.

In 1843, Ascension received perhaps its most consequential visitor in its million-year history—at least as far as the island's flora was concerned.

British botanist and explorer Sir Joseph Hooker visited the island during his return from a four-year voyage to Antarctica with explorer James Clark Ross. After observing the plant life on the island, he concluded that this smattering of a few species of plants was inadequate, and that something must be done to remedy the situation.

When Hooker returned to England, he was invited by the British Admiralty, which oversaw the occupation of Ascension, to offer suggestions on how to improve the productivity of the island's vegetation and conserve water from its limited rainfall. He reasoned that both goals could be achieved by introducing new plant species to the island. For example, he believed that water droplets could condense on the leaves of trees planted at the highest elevations, which would then drop and moisten the ground. Furthermore, increasing the cover of plants on the hillsides could help create soils and reduce evaporation. Accordingly, Hooker recommended that new species of plants be transported from around the world and planted on Ascension. He made a list of good candidate species, which could be shipped in from nearby Saint Helena or farther afield from botanical gardens in England, Sydney, and Cape Town.

The Admiralty agreed to Hooker's plan, and in 1847, they sent a gardener to Ascension to direct the process of establishing new species on the island. For the next few decades, seeds and seedlings from around the world were sent to the island. Workers planted eucalyptus trees, rubber trees, Norfolk Island pines, bamboo, agave, prickly pear, and orange trees, to name a few. One shipment in 1858 included 228 species of plants. All told, hundreds, and perhaps even thousands, of species were sent to Ascension Island, and although many did not survive the journey or did not thrive in the island's climate, some took root. In fact, so many introductions were successful that the new plants began to transform the island, especially in the higher elevations where rain and fog were most plentiful.

In 1958, Eric Duffey led a systematic survey of the plants of Ascension, a century after the heyday of the deliberate introductions. He identified 128 species that showed evidence of having naturalized, whereby they were propagating themselves. In 2013, botanists estimated that from 200 to 300 plant species were growing on Green Mountain. In just 150 years, the plant diversity of the island had increased at least tenfold.

A CLOUDED JUDGEMENT

Imagine that you're sitting beside a calm forest pool, surrounded by verdant plant life. At the edge of the pool, fern fronds cascade into the water. The temperature is 72°F (22°C), and the air is extremely humid, so that every surface is glistens with moisture. Behind the ferns, giant leaves of tropical plants overhang the pool, soaking up some of the sunlight that breaks through. High above you is a living rattan wall of green-leaved bamboo that looks impossible to penetrate. This is the top of Green Mountain today.

The plant community on Green Mountain is now considered a cloud forest, which is a special kind of ecosystem that collects most of its moisture from the surrounding humid air rather than through rainfall. It's a far cry from the simple "carpet of ferns and here and there a shrub" that Hooker described when he visited the top of Green Mountain in 1843. Further down the mountain are tall trees, covered by a green carpet of mosses and epiphytes, with a rich understory of flowering plants. Beyond the dense growth is a verdant landscape that extends far down the mountainside, ending in desert-like lowlands.

Cloud forests are rare globally, and they typically occupy small pockets in tropical mountains such as those in Costa Rica, Indonesia, India, Hawai'i, and Papua New Guinea. Standing within such a forest, one might get the impression that it has been there for time immemorial, with all the plants and animals slowly evolving together to produce a forest that drinks in the air to support a community of life. Some ecologists have proposed that this is how ecological communities form: through eons of interactions, species gradually develop separate niches that collectively produce the emergent patterns of the ecosystem.

Evidence that species have slowly evolved together is quite easy to find in nature. For example, consider an orchid in Madagascar that has coevolved with a hawk moth: the orchid has an extraordinary foot-long nectar tube that can be reached only by its coevolutionary partner, which has a foot-long tongue. Or consider the tiny black-and-white Caribbean shrimp that have evolved to look exactly like the spines of their companion sea urchins, to keep them well hidden from predators. Or consider ants in Africa that have developed a mutualistic relationship with acacia trees: the ants get a place to live and food to eat, while the tree gets an army of ants that will defend it from pests.

Such relationships take evolutionary time to form, as each successive generation becomes better adapted to survive in relation to other species. These interactions between individual species may lead us to conclude that entire ecosystems are similarly built, with each species evolving to fill a niche, collectively giving rise to complex emergent patterns, such as those found in a cloud forest. However, this idea about how an ecosystem forms is largely incorrect, because species can quickly arrange themselves into new ecosystems, and the plants of Ascension Island show us how this happens.

Ascension's cloud forest formed within a century, as plant species collected by humans around the world were thrust together into an ecological melting pot. At first, the system would have changed very rapidly, as each new species struggled to establish itself in this new place. Each would have relied on phenotypic rescue to adjust its physiology or appearance to cope with local conditions, and then reproductive rescue to grow its population in uncrowded conditions. Many species faced a losing battle, and they never gained a foothold on the island. But for several hundred other species, the rescue effect was strong enough that they successfully adjusted and fended for themselves. Now they have self-sustaining populations that have formed an entirely new kind of environment. Ascension's cloud forest did not require eons of evolution to form; instead, it formed quickly by drawing on the rescue effect to take advantage of the zealous efforts of a few people to increase the plant diversity of the island.

In ecological terms, Ascension's plants constitute a *novel ecosystem* that comprises a mixture of species that has never before existed. Conservationists have often pooh-poohed novel ecosystems, because they are viewed as being less intrinsically valuable than ecosystems comprising native species, which more closely resemble the ecosystems of the past.

The views of native species as inherently good and recently introduced species as inherently bad are demonstrated in a 2009 management plan for Ascension's few remaining native plants. For example, the authors note, "The consequences of anthropogenic change have been severe," as areas are "infested with undesirable species," and habitats for native species are "very severely degraded." In contrast, areas that still support native species "have a high biodiversity value." This value assessment puts an overwhelming thumb on the scale in favor of the native species mix that some people believe should be present, over non-native species that have "infested" the island.

Without a doubt, the current plant community on Green Mountain is a far cry from its historical roots, with only a few native species remaining within the globally sourced botanical gardens run wild. However, the new ecosystem offers many assets that were not present in the preceding version, such as shade, timber, soil generation, food, and high biodiversity—and the jury is still out regarding whether it provides more surface water. If this same group of species had populated Ascension without the help of people, some might hail the area as a globally significant cloud forest that requires stringent protections. But should its value be discounted because people played a role in assembling the species?

At the end of the day, very few of us have a tangible stake in the plant community on Green Mountain. Most of the species that grew on the island prior to human intervention still remain, though some are at risk of extinction without ongoing care and maintenance by island residents and naturalists. Reasonable people might disagree about the management goals that should be pursued for Green Mountain going forward. Some might argue that all available resources should be brought to bear to try to eradicate the invasives in favor of restoring the natives. Others might argue that the new species should be embraced, regardless of the endangered native plants. The 2009 management plan for Ascension's native plants advocates for a middle ground: prioritize the protection of native plants in the few places that are still dominated by them, while creating novel habitats in which native species can persist among the more recent arrivals.

Although the questions about how to manage Ascension's plant community today are important to consider, the greater global value of the story of the island's cloud forest may be in demonstrating how a mishmash of new species was easily assembled into an entirely new ecosystem. This is good news for conservation, because we are entering an era in which nearly all ecosystems are experiencing changes in species composition. Green Mountain's cloud forest shows that newly introduced species can quickly form complex ecosystems. We should expect other ecosystems to reorganize similarly, while continuing to provide benefits to other organisms and people.

The truth is, ecosystems never arrive at some sort of stable equilibrium and then remain there. Recall the changes in the distribution of life at the end of the Pleistocene, or the sockeye salmon colonizing Bristol Bay, or the more recent drying and refilling of Lake Victoria that started the evolution of the current haplochromine species flock. These events all occurred within the last 20,000 years or so—a drop in the bucket in evolutionary time. With a longer frame of reference, all these ecosystems could be considered novel prior to their more recent, sometimes human-induced, changes. In fact, the term *novel ecosystem* may be redundant, because it implies that there's something fundamentally different about an ecosystem that has recently shifted its composition.

The next phase of life on Earth will be characterized by more waves of novelty as people drive species to new places and the environmental conditions that govern ecosystems shift because of climate change and other human activities. People get to decide for themselves how to view this novelty. One option, which author Fred Pearce calls "green xenophobia," is to revile new species and novel ecosystems, while valuing only the familiar species and ecosystems of the past. But perhaps green xenophobia is the result of clouded judgement, which causes us to devalue novelty unreasonably. As an alternative, people can choose to recognize the value of novelty, which highlights the ability of species to adapt to new conditions and reorganize into something new, all the while providing services to people and wildlife. For Ascension Island, embracing novelty doesn't mean that conservationists cannot work to save its native plants or mourn those that may be lost. It means viewing the cloud forest on Green Mountain as something remarkable, with its own merit and value.

ECOLOGICAL ASCENSION

Although the exact events are lost to history, some believe that Ascension Island was named by Afonso de Albuquerque, a Portuguese general and admiral, who reportedly sighted it on Ascension Day in 1503. It seems fitting that the island would be named in honor of the Christian tradition of Jesus Christ rising from his mortal death into heaven. At its core, the Christian Ascension story is about rebirth and transition to the next stage of existence. And this is what has happened on Green Mountain, where more than a century of disruption has unambiguously brought about the demise of its prior ecosystem. But from this death, there has been rebirth, as a new ecosystem has ascended into the clouds. In this, the Ascension Island ecosystem will not be alone, because the rescue effect drives species to find ways to live in new climates, new places, and among totally new combinations of life on Earth. Perhaps it's time for people who love nature to start seeing this novelty for what it is: a compelling demonstration of how nature transforms itself in response to a changing world.

CHAPTER 9

A Bright Future

NATURE IS AN ART MUSEUM, with living organisms and ecosystems akin to great masterpieces that should be protected in their original state in perpetuity for the benefit of humanity. For more than a century, this idea has been central to conservation biology, whereby the goal of conservation is often to keep nature from changing or to restore it to some past state.

The concept of nature as a museum underlies the most pervasive tool for protecting wildlife: the nature park, a place deliberately set aside to protect it from many direct human actions. The typical goal of a nature park is to maintain an area in its historical state, which is usually related to a certain group of desired species (almost always native species) in a certain abundance. Accordingly, visitors are not allowed to remove anything from many parks, be it an animal, a tree, a rock, or a feather. To ensure that parks do not change, park managers attempt to keep out or remove invasive species and reduce threats such as wildfire. And if all else fails, parks can be restored to the desired mix of species using approaches such as cultivating and transplanting target species of plants (boosting reproductive rescue) or reestablishing lost species such as wolves or brown bears (boosting demographic rescue).

Embedded within this school of thought is the idea that the greatest value of a wild ecosystem is achieved when it is in its original state, with an unchanging group of species whose populations remain stable over time. Similarly, a goal of museums is to preserve individual artworks in their original states. Nobody wants to see changes to the original *Venus de Milo* or Vincent van Gough's *The Starry Night*—they should look as close to how they looked when they were first created, with as much fidelity as possible, and they should be shielded from dirty hands, UV light, air pollution, and subsequent generations of artists who might wish to tinker with the original. By applying the same kind of value to nature, it makes sense to protect it similarly: minimize human impacts, maintain species populations, save the oldest and largest trees, root out invasive species, and carefully restore damaged ecosystems to the states they previously held.

Unfortunately, even for those who champion the idea of maintaining nature as a museum, this won't work, because the idea is built on a faulty assumption: that nature stays the same over time when people are not messing with it. We know this is not true from countless data-points, such as the fossil record, which shows that life has continued to evolve and change for billions of years—from an era of microbes, to the rise of plants, through the many different ages of dinosaurs, and finally to the explosions of insect, bird, and mammal diversity that we see today.

Ecosystems change at timescales of thousands of years too. The earlier chapters of this book are chockfull of examples of ecosystems reorganizing into something new in response to changes in the environment. Since the end of the Pleistocene, American chestnuts and many other tree species have spread north into New England, sockeye salmon have populated Bristol Bay, and the steppe ecosystem has given way to forests in Western Europe. Each of these examples resulted in fundamental changes to local ecosystems that did not involve human meddling. Change is also constantly underway at much shorter timescales, as local environments shift from one decade to the next, some species move to new locations or go extinct, and other species evolve new capabilities. Stasis isn't the norm in nature; change is.

The nature-as-museum conservation model is also limiting because it requires humans to articulate what the desired state should be. Because many ecosystems are constantly in flux, this arbitrary choice becomes important. Consider, for example, a Caribbean coral reef: What should be its baseline for conservation? Perhaps we should set conservation goals based on the reefs of the 1970s, before significant coral declines, when reef-building corals were more abundant. Restoring most Caribbean reefs to their 1970s state would require heavily bolstering existing coral populations, which may not be successful given contemporary problems such as coral diseases and increasing water temperatures.

But even if we could achieve this, is it really the right goal? The reefs of the 1970s had already been heavily impacted by humans, especially through fishing. Prior to the colonization of the Caribbean by Europeans, the area would have had a much higher abundance of large animals such as sharks, crocodiles, and groupers. In the absence of these large predators, the reefs of the 1970s were probably overrun with smaller fish species that flourished in their absence. These reefs may have been beautiful and full of life, but they were certainly not beyond the reach of humans.

Perhaps instead of setting the management target to the 1970s, an era of abnormally low large predator abundances, the target for a Caribbean reef should be based on what existed a few hundred years earlier, when more large predators ruled the reef. Some, such as sharks and groupers, could return to the area if fishing were restricted. However, fishing restrictions are often politically and practically difficult to achieve, particularly when people rely on the coastal ocean for food. Other species, such as crocodiles, could be reintroduced where they have been extirpated, though many people would oppose returning these dangerous animals to coastal areas.

If we were to set the restoration of large predators as a management goal, we could use the *Jardines de la Reina* (Gardens of the Queen) in Cuba as a model, because fishing on these reefs has been heavily restricted across a large area for a long time. I visited these reefs in 2016 and was blown away by the abundance of large animals. Everywhere I looked, I saw sharks, giant groupers, tarpon, and snappers— even a healthy population of crocodiles live in the mangrove forests near the reefs.

At first, I thought that the 2016 Jardines reefs must be similar to the pristine reefs that existed before people changed them. But I soon realized my initial impression was wrong. Yes, these Cuban reefs include an incredible population of large predators, but they are experiencing the same kinds of coral declines and coastal pollution issues experienced by other reefs in the region. Furthermore, the Jardines de la Reina reefs, like all Caribbean reefs, are missing the Caribbean monk seal, which went extinct in the mid-twentieth century. Perhaps the monk seal was very important to Caribbean reef ecosystems prior to being hunted to extinction by humans; we don't really know. We do know, however, that the full complement of predators on Caribbean reefs cannot be re-created without either resurrecting the monk seals or introducing similar animals, such as the endangered Hawaiian monk seal. Ultimately, even a remote, well-protected reef like the Jardines de la Reina could never be fully restored to a Caribbean reef of 500 years ago. Instead, it's something new. It's awe-inspiring for sure, but it's not an antique masterpiece in an art museum.

As with all ecosystems, we have no objective way of choosing a baseline for what Caribbean reefs should be like, and even if we did, restoring reefs to this state is probably impossible. As a result, decisions about baselines necessarily boil down to human values, and decisions about what should be considered important and what should not. But as soon as those values settle on an arbitrary historical target, they will be destined to fail, because too much has already changed in species distributions, climate, and other environmental conditions. More importantly, the state of an ecosystem is inherently ephemeral, so past ecosystems can never truly be restored, nor can their present states be entirely shielded from ongoing and future change. Aiming to do so is ultimately folly, and because the resources available for conservation are always limited, trying too hard to prevent change or to restore the past will likely siphon resources from other conservation approaches that may provide more benefits.

So if the nature-as-museum model is ultimately destined to fail, how should people approach conservation? The answer lies in the rescue effect itself, which is a process rather than a state. By managing to promote the rescue effect, the goal shifts from trying to prevent change to guiding change in directions that people value. This is an emerging paradigm in conservation, and it's less like managing an ancient art museum and more like parenting.

PARENTING LIFE

Although people approach parenting differently, I'd argue that one important goal is to equip children with the ability to take care of themselves. To do that, parents must choose when and how to intervene in their children's lives. A lot of intervention is required when children are young and vulnerable. But even at that stage of a child's life, every parent knows that this obligation is probably a temporary one, and the child's process of growth and development will inexorably move forward. The goal of a parent is not to stop change, no matter how much they may want to hold onto a child's first giggles or steps forever; instead, one of the responsibilities of parenting is to help children pass to the next stage. As children develop, parents can decide when and how to intervene in their lives by weighing the costs and benefits of doing so. But parents don't get to decide everything; children will navigate their lives based on their own experiences, opportunities, and desires. And this is how we should view nature. We should celebrate the fact that nature changes over time, fondly remember what nature was like in the past without being tied to it, and carefully choose whether and how to intervene to shape the natural world of the future.

There's no question that this view of nature is paternalistic, but it's the only way people can successfully manage nature on a planet dominated by the changes we are creating. Ultimately, nature doesn't care how many species go extinct, and it doesn't care how much we change the climate. Nature can adapt to all these changes on its own, as it's been doing for billions of years. That said, the ways that nature adapts may not always be to our liking, especially if we lose species or ecosystem services that we once valued. To protect our own interests, then, we must step up to our parental role and intelligently choose whether and how to intervene as nature inevitably changes. Doing this will build on much of what the conservation movement has already learned about rescuing species and managing ecosystems (largely under the nature-as-museum model), while also considering whether and how to apply new tools. But the most important shift in conservation needs to be in the goalpost: we must move it from keeping nature the same to helping nature change. To achieve this goal successfully, we will need to focus on three things: boosting the rescue effect, slowing the rate of change, and celebrating novelty.

Boosting the rescue effect

Viewing nature through the lens of the rescue effect forces people to consider the different processes that automatically help species adapt to change. It also helps clarify the full range of options that we have to intervene on behalf of species that are struggling.

Evidence for all the processes that make up the rescue effect can be seen everywhere in nature. For many species, phenotypic rescue—changes in behavior, appearance, and physiology—will be enough to help them adapt. For others, such as mountain pygmy-possums and corals, phenotypic rescue won't be enough, and they'll need to rely on other adaptive processes to survive. When populations are reduced by some disturbance, such as the salmon fisheries in Bristol Bay or tiger poaching in India, reproductive rescue will often bring their numbers back up automatically. Small populations can also be rescued by the immigration of new individuals, which leads to demographic rescue in connected populations of species, such as coral larvae that can drift from one reef to another. When populations decline to the point that inbreeding causes genetic problems, immigration of individuals, like Panna Lal walking from one park to another, can add much needed genetic diversity, leading to genetic rescue. If environmental conditions become too harsh for any of these processes to succeed, a species can also undergo evolutionary rescue, where evolution favors a new set of genetic combinations—and this already appears to be happening for species like Lake Victoria's haplochromine cichlids. And, finally, if a species' preferred environmental conditions are no longer available in their current homes, many organisms will move to new locations via geographic rescue, such as the sockeye salmon that may be expanding into the Arctic. Collectively, all six of these processes, working on their own or in complex combinations, give rise to the rescue effect: the innate and automatic tendency for species to adapt to a changing world.

We do not have to rely on the rescue effect occurring entirely on its own. In fact, we can choose to boost each of these processes for species that are struggling to adapt to environmental change. Phenotypic rescue can be boosted indirectly by reducing stressors for species such as mountain pygmy-possums and corals. Reproductive rescue can be boosted by reducing mortality rates for species such as the haplochromine cichlids that are being depleted by Nile perch predation, or by breeding new corals to transplant on reefs. Demographic rescue can be boosted by moving individuals to populations that have declined or disappeared, such as the tigers being reintroduced to Panna Tiger Reserve. Genetic rescue can be boosted by adding some genetically diverse individuals to small populations that have become inbred, such as the captive-bred mountain pygmy-possums released on Mount Buller. Evolutionary rescue can be boosted for species that are failing to adapt by selectively breeding or genetically engineering species such as the American chestnut or Caribbean corals. Geographic rescue can be boosted by moving species like mountain pygmy-possums to new locations, where they may thrive in the future. And finally, we may someday master resurrective rescue, which will enable us to create facsimiles of species that have already gone extinct.

In a world that is rapidly changing, both action and inaction can involve significant risks. Failing to intervene in new ways may be tantamount to accepting negative outcomes, such as extinction, while pursuing interventions may introduce new risks that change an environment for the worse. How will we navigate such difficult decisions? Should we move species such as the mountain pygmy-possum to new places? Should we genetically engineer species such as the American chestnut to save them? Should we introduce new species of corals to reefs where the local species have declined? Should we resurrect extinct species such as the mammoth or the cave bear? Should we cut down the cloud forests on Ascension Island to help the native plants that are struggling to persist?

Ultimately, there are no correct answers to any of these questions, only judgement calls to be made based on our assessments of the risks and values. What is clear is that nature's rescue effect is extremely powerful on its own—indeed, it has allowed life to persist for billions of years—and it's even more powerful when we intervene.

That said, for some species, the world will change so quickly that they will be lost no matter what we do to boost the rescue effect. To prevent this from happening, our best overall strategy is to slow the rate at which the world is changing.

Slowing the rate of change

Years ago, I asked one of the world's foremost experts on coral evolution whether any places in the ocean were too hot for corals to live prior to human-caused climate change. After a pause, he replied that he couldn't think of any. Many parts of the ocean are too cold for corals, but they've flourished even in the hottest reefs in the Middle East and in shallow lagoons that bake in the tropical sun. Why, then, are corals struggling to survive in today's rising ocean temperatures? The difference is the span of time over which the temperatures rose.

The corals that adapted to high temperatures in the past had thousands of years to adjust their phenotypes and to relocate, evolve, and shuffle their symbionts to find combinations that worked. The corals that are affected by climate change today have not had the luxury of time to adapt, with grim results. I have no doubt, however, that corals could adapt to higher temperatures, just as I have no doubt that mountain pygmy-possums could adapt to live in warmer climates, or that new species of Lake Victoria cichlids could arise. But with the world changing so quickly, will they have enough time?

Conservationists have long recognized that many species struggle when their environment changes too quickly, and this is why so many conservation campaigns focus on reducing the drivers of change. Much of this work is locally driven, where conservation advocates try to reduce new sources of change. While locally based work is promising, the biggest source of change for many species and ecosystems going forward is likely to be global changes in climate. As we release more and more greenhouse gases into the atmosphere, we are changing every ecosystem on Earth, via new patterns of temperature, rainfall, storms, wildfires, currents, and chemistry, to name a few.

Today, climate change is becoming more of a political and economic problem rather than an engineering problem. We have many great, and constantly improving, technologies at our disposal to help us reduce production of greenhouse gases and slow climate change—if we are willing to pay for their implementation. Climate change is a problem that we must solve, whether for our own wellbeing or for the wellbeing of other living organisms. The good news is that many species are already adapting, and most have scope for adapting further, giving us time to act. However, my largely optimistic take on the future of life on Earth is predicated on humanity showing the resolve to address climate change in the coming decades; otherwise, the pace of change will simply overcome many species' ability to cope with it, and a lot of the things people value about nature will be lost.

Celebrating novelty

I'm an amateur underwater photographer. Whenever I travel to places with coral reefs, I schlep an extra suitcase full of gear that I use to try to capture some of the awe and beauty that I am privileged to witness while diving. Some of my favorite photographic subjects in the Indian and Pacific oceans are the lionfish—several species of strikingly colored and patterned fish. Lionfish have unusually long fins that they use to corral unsuspecting small fish into corners, where they are no match for the lionfish's suction-powered jaws. Lionfish also have venom-filled spines in their fins that act like hypodermic needles when touched. If a big grouper tries to swallow a lionfish, it will get a mouthful of burning stings.

In the late twentieth century, lionfish were accidentally introduced to Florida, and they have since spread throughout the Caribbean and neighboring waters. Many communities have been alarmed by the arrival of these voracious new predators and see them as a threat to their reefs. Researchers have confirmed that lionfish can change the reef by consuming a large number of small fish, reducing their populations. Many people have thus branded the lionfish public enemy number one on Caribbean reefs and are taking steps to eradicate them, including issuing special permits to spear them in ways that are not permitted for other fish. To encourage consumption and hunting of lionfish, some communities have even sponsored lionfish cook-offs.

When I dive in the Caribbean, I often see lionfish hiding in the cracks and crevices of the reefs, which, in the past, caused a small—yet telling—dilemma for me. Should I take pictures of lionfish to show off their striking beauty? On the one hand, conservation orthodoxy encourages me to revile these fish: I should kill them for the harm they cause. They don't belong here and should be eradicated. They're worse than having no value—they have negative value. On the other hand, they are truly a pinnacle of natural selection, with their elegant fins and formidable defenses, and they've managed to adjust to life in an entirely new ocean. My internal conflict was really about my own contradictory view about novelty: I was excited to see these amazing animals, but at the same time, I felt that they didn't belong.

Today, I'm beginning to appreciate lionfish in a different light, and I embrace them as part of modern Caribbean coral reefs. One reason for this change of heart is purely pragmatic: I know that, with existing technology, we cannot remove all the lionfish from the Caribbean, just as we cannot eradicate chestnut blight from the United States or remove Nile perch from Lake Victoria. Although introduced species can sometimes be completely removed (especially on small islands), there's usually no practical way to do this, even if people decide they want them gone.

But why settle for a grudging conclusion that eradication is an impossibility, when we could instead admire these newcomers? Admiring lionfish does not require a denial of the fact that they eat other fish and decrease their populations, but it does open the possibility of seeing them as a legitimate member of an increasingly novel ecosystem. In fact, lionfish are just one small part of the novel ecosystems forming on Caribbean reefs, with fewer large predators and reef-building corals and more sponges and soft corals. Perhaps this entirely new ecosystem deserves admiration in its own right? After all, Caribbean reefs have rapidly reorganized in the face of a changing world into a system that still provides many benefits to people and to ocean creatures, even if the exact nature of those benefits has changed in recent years. Today, I consider lionfish a valid part of that new ecosystem, and going forward, they always will be.

Caribbean reefs epitomize the future of ecosystems. Even if we don't accidentally or purposefully introduce species that change their environments, organisms are finding their own ways into new places all over the word. And at the same time, species that have long resided in one place are undergoing their own changes—perhaps moving elsewhere, taking on new ecological roles, or even evolving into something new. The net result will be novelty everywhere.

To be successful in conservation in the Anthropocene, we need to stop shunning novelty and celebrate it instead. Novelty shows the rescue effect in action, whether it is a spontaneously formed cloud forest on Ascension Island or a newly emerging cichlid species in Lake Victoria. Novelty is proof of the tenacity of life and its potential for the future. Novelty—rather than a hermetically sealed museum—is success.

A GLASS MOSTLY FULL

Nature enthusiasts sometimes overestimate the immediacy of the risks to life on Earth, creating an atmosphere of gloom and doom. For example, I was recently speaking with a friend about how climate change and other human-caused change might affect living species. He was highly concerned about the future of species, so I asked him, "How many species do you think are at near-term risk of extinction?" After acknowledging his uncertainty, he offered an estimate of 90 percent, based on his subjective perception of the seriousness of climate change. Although 90 percent is an unreasonably high figure, it's consistent with the sense of dread that exists among some people who get caught up in the bad news.

How many species actually are at risk today? In the introduction to this book, I noted that based on one estimate of the current rate of species extinctions, we stand to lose about 1 percent of our current species within the next century. Although that rate is extremely high compared to extinction rates prior to rapid human population growth, it's hardly an immediate existential threat to the future of life. The rate also indicates that the rescue effect has worked for nearly every species on Earth, as most of them have already successfully survived the changes created by humans. But what will happen going forward? The greatest concern is that the extinction rate will continue to rise, steadily eroding the diversity of life.

To get a sense of how extinctions may play out in the future, we can look at species one by one to assess their individual risk of future extinction. The International Union for Conservation of Nature (IUCN) has developed a system for measuring extinction risk, which has been applied to thousands of species globally. When a species is discovered to be at heightened risk of extinction, it is added to IUCN's Red List, which is the world's most comprehensive database of its kind. As of March 2022, the Red List documented more than 40,000 species that are threatened with extinction globally. This is a huge number of threatened species, which includes many globally iconic animals such as pandas, gorillas, and blue whales. Furthermore, about 8000 of these threatened species are in the critically endangered subcategory, which, according to IUCN, means they are "facing an extremely high risk of extinction in the wild."

All told, these 40,000 threatened species constitute about 2 percent of the world's two million known species, while the subset of species designated as critically endangered is about 0.4 percent. At first blush, these numbers are refreshingly low. Based on these numbers alone, one might conclude that 98 percent of Earth's species are doing just fine, while 99.6 percent of our species are probably not facing an immediate risk of extinction. Assessing future risk is not that straightforward, however. Although some groups of species, such as mammals, have been particularly well assessed, the vast majority of the world's species haven't been through the IUCN Red List assessment process. And because we know that more at-risk species will be identified in the future, these percentages of threatened and critically endangered species will grow.

The IUCN estimates that a whopping 26 percent of all mammals are threatened with extinction, perhaps because mammals are particularly sensitive to human actions. For instance, many mammals are hunted for economic gain or killed by humans who feel threatened by them, which is one of the main reasons that rhinos, tigers, and elephants face heightened extinction risks today. Some mammals, such as orangutans, are vulnerable particularly because they need large amounts of undisturbed habitat in order to survive, and appropriate habitat is becoming a rarer commodity as human land use continues to grow. Still others, such as mountain pygmy-possums, are confined to small geographic areas in rapidly changing environments.

Losing all of the mammals on the Red List would be a terrible blow to the diversity of life as we know it, forever changing ecosystems. But is this bleak future going to occur anytime soon? Absolutely not. If we look a little closer at the classifications, it's clear that many mammals included on the Red List are not at high risk of immediate extinction. This seems like a paradox: How can so many species be included on the Red List yet not be immediately threatened with extinction? The answer is found in the evaluation criteria used, which enables many species to be designated as threatened long before they are in imminent danger of extinction.

For example, 42 percent of the threatened mammals on the Red List are subcategorized as vulnerable. This designation can serve as an important early warning system, but it may or may not indicate that an animal is heading rapidly toward extinction. To be designated as vulnerable by the IUCN, any one of six different warning signs needs to be documented. For example, if the estimated extinction risk can be shown to be greater than 10 percent over the next century, the species is classified as vulnerable. Putting that into perspective, if we followed ten vulnerable species with a 10 percent chance of extinction, we would expect that only one of those species would go extinct after 100 years.

Other warning signs include measures of whether and how much a species' population is declining, whether the species is confined to a small geographic area, whether it has a small population size, or how much its population tends to fluctuate over time. All of these measures are correlated with higher extinction risks, so it makes sense to call attention to them, but none of them on its own necessarily means that extinction is imminent. Instead, many species designated as vulnerable are probably undergoing an adjustment to a new set of conditions, and the rescue effect will help ensure that they persist for the foreseeable future. Moreover, if the species takes a turn for the worse, people have time to actively improve their chances of persistence.

The same cannot be said for the species that are designated as critically endangered, however. These species would be likely to disappear soon without intervention. Fortunately, only about 3.5 percent of mammals are designated as critically endangered, and many of these species can be helped with concerted efforts to boost the rescue effect. Based on IUCN's data, conservation efforts are largely holding the line for mammals so far, with most species managing to persist presently, and their overall risk levels have remained fairly stable in recent decades, rather than rapidly increasing, as the world keeps changing.

Comprehensive data for other species groups shows a variety of patterns in terms of extinction risk. Like mammals, bird populations are by and large holding steady. The overall risks are slightly increasing for amphibians. For corals, the risk outlook has been deteriorating rapidly, as coral bleaching and other threats decimate their populations. But the outlook for corals may not be as dire as it seems, because we have increasing reason to believe that corals may be able to keep up with climate change through mechanisms such as evolutionary rescue, provided people begin to reduce greenhouse gas emissions in coming decades.

Taken together, what do these trends mean for the future of life on Earth as we know it? Many conservationists take a glass-half-empty view, but the data on extinction risks points to an alternative answer: the glass is still mostly full. Because of the rescue effect, combined with help from people, the vast majority of species are so far successfully adapting to a changing world. Moreover, a low proportion of species are facing imminent extinction risks, even within particularly beleaguered groups such as mammals. The low percentages of species that are critically endangered, combined with our growing set of tools for helping species that are struggling, suggests that we are not yet locked into a future in which a large fraction of Earth's species are lost to extinction. Instead, if we choose to, we have the time and the tools we need to shape the future in ways that will help most species pass through the current period of rapid change.

THE GREATEST RESHUFFLING

This book has focused on how the rescue effect is helping species and how people can choose to assist these processes. But let's consider another dimension: how the rescue effect will influence life in the coming millennia.

The Lake Victoria haplochromine cichlids give us a glimpse into how Earth's future might look, with species regrouping under novel conditions and then starting to evolve into new species. Although the haplochromines are in a class of their own in terms of their ability to evolve quickly, the same kinds of responses will play out over longer timescales for untold numbers of species as natural selection pushes them in new directions.

Consider, for example, the future of housecats. Housecats have been introduced to all the continents (except Antarctica) and hundreds of islands, thanks to people, and they are frequently included among the most destructive invasive species on the planet. Within thousands of new populations thriving in a variety of climates and with combinations of different species, housecats are surely already adapting to local conditions. Imagine that you could take a time machine a million years into the future to see how housecats have adapted. Left to their own devices, many populations will likely have evolved into separate species. Perhaps you'd find furry descendants of Bengal housecats in the Himalaya Mountains, giant swamp cats in northern Australia, tawny lizard-eating cats in the Mojave Desert, and striped tree cats in the jungles of Indonesia. Hundreds of islands would have their own endemic cat species emerging; the cats in Cuba would be different from the cats in Hispaniola, which would be different from the cats in Puerto Rico or Jamaica. In total, the world's biodiversity of cats would be increasing, with hundreds of new species.

If you value nature as a museum, this is a terrible outcome. Housecats are already increasing the risk of extinctions among their prey, and they are contributing to the restructuring of their new ecosystems. But if you value how the rescue effect helps life adapt, housecats offer hope. Their novel ecosystems will include many species—some that have survived in place for a long time, some that have arrived more recently, and others that are evolving into something new.

By transforming habitats, moving species to new locations, and altering the climate, people are engineering the greatest reshuffling of life ever experienced on Earth. Although all this change is creating new challenges for many species, it is also creating new opportunities. Indeed, over time, we are setting housecats, and thousands of other species, on a path toward unprecedented diversification. In a million years, there may be scores of new plant species on Ascension Island; hundreds of new cichlids in Lake Victoria, including types that are even more wondrous than the paedophages and scale-eaters of the past; new salmon species in the Arctic as well as the places where salmon have been introduced, such as New Zealand or the North American Great Lakes; new corals and lionfish in the Caribbean; and perhaps even new chestnuts emerging from their hybrid and genetically engineered ancestors. Indeed, if humanity can manage to get a handle on greenhouse gases and otherwise use our technological prowess to lessen our footprint on the planet in the decades to come, we should expect the next age of life on Earth to become more diverse than any before, as the species that we reshuffled evolve to thrive in a new world.

Ultimately, imagining the future of life on Earth doesn't have to be limited to a depressing list of the organisms that we could lose. True, we have good reason to be concerned about how we are affecting species and ecosystems. But even as we acknowledge the risks to species and ecosystems, fight to reduce them, and mourn our losses, we can admire how most species are successfully adapting and creating new ecosystems. We can even recognize that the future of life may be more exciting than ever before, as it gives rise to new species and novel ecosystems. For that, we should revere the rescue effect—which will always be there, helping life adapt no matter how the world changes. At the same time, we have a growing ability to shape our future by slowing the rate of environmental change while carefully choosing whether and how to boost the rescue effect for those species that we value the most. Because of the rescue effect and our ability to strengthen it, we have good reason to believe that the future of life on Earth will be bright.

Notes

INTRODUCTION

Page 10, what I call *the rescue effect*: Brown and Kodric-Brown (1977) first used the phrase "the rescue effect" to describe how immigration can reduce the rate of extinction in small populations. Today, the phenomenon described by Brown and Kodric-Brown is known as *demographic rescue*.

Page 10, the rescue effect automatically turns on when a population is stressed or declining: Some of the processes that make up the rescue effect are truly self-regulating. For example, reproductive rescue is part of a feedback mechanism that helps small populations grow and causes large populations to shrink, which tends to keep populations somewhere in the middle. In contrast, immigration and emigration can occur by chance, in which case the rescue effect is driven by serendipity rather than a self-regulating feedback mechanism.

Page 12, more than 99 percent of the species that have ever lived are now extinct: This estimate (Jablonski 2004) is largely a function of time: life on Earth is so old that there has been enough time for species to arise and disappear many times. Those species that exist today represent the lineages that have managed to persist through this process.

Page 13, mass extinction was caused by a giant meteor: Schulte et al. (2010) describes evidence that an asteroid impact about sixty-six million years ago caused a mass extinction that included dinosaurs.

Page 13, shook the Earth with the equivalent force of a magnitude 11 earthquake: Day and Maslin (2005).

Page 13, enough to trigger aftershocks and volcanos worldwide: Richards et al. (2015) used mathematical models to conclude that the impact of the asteroid was likely strong enough to trigger volcanic eruptions.

Page 13, creating something akin to a nuclear winter: Vellekoop et al. (2014) argued that, in the months to decades following the asteroid impact, the Earth experienced an "impact winter" that "was likely a major driver of mass extinction because of the resulting global decimation of marine and continental photosynthesis."

Page 14, extinction rates today are 1000 times higher: De Vos et al. (2015) estimated a background rate of extinction of 0.1 extinctions per million species per year and noted that current extinction rates are roughly 1000 times higher than natural background rates.

Page 14, about one species would go extinct every five years: Scientists have described about two million currently living species, though there may be many millions more yet to be discovered. Considering the two million described species, the background rate of extinction is about one species extinction globally every five years. Because our current rate of extinction is roughly 1000 times higher, we're now expected to lose roughly 200 species per year.

CHAPTER 1

Page 21, an informal team of photographers and wildlife managers had solved the mystery: An account of the tiger's identification is provided at https://newsroom24x7.com/2017/06/12/panna-tiger-reintroduction-project-and-bahubali-2-the-story-of-the-panna-tigers/.

Page 21, He was thereafter given the fitting name of Panna Lal (Hindi for son of Panna): In my research, I have found many other names for Panna Lal. In Panna Tiger Reserve he was given the designation P213(21), a code used to denote his ancestry, and he was referred to as Bahubali 2, or Baahubali 2 (Sanskrit for one with strong arms), in some newspaper articles when he first moved from Panna to Bandhavgarh. He was given the designation T71 once he entered Bandhavgarh, and finally Panna Lal, which some shorten to Panna. In this chapter, I refer to him as Panna Lal.

Page 21, the challenger may start his reign by killing all his predecessor's cubs: Singh et al. (2014) described the practice of male tigers killing the cubs of rivals, as well as ways that tigresses try to avoid this fate for their cubs.

Page 23, Scientists have used DNA samples to identify nine distinct tiger populations: Luo et al. (2010) describes how genetic information has been used to identify genetically separated populations of tigers.

Page 24, tigers have lost 93 percent of their historic territory: Dinerstein et al. (2007).

Page 24, There is a market for nearly every part of a tiger: Moyle (2009) describes the black market for tiger products.

Page 25, The international trade of tiger products is banned: Information on India joining CITES is located at https://cites.org/eng/disc/parties/chronolo.php.

Page 25, the global wild tiger population has crashed: Seidensticker et al. (2010) estimated that 4500 tigers were living in the wild in 2008.

Page 26, India's twenty-second tiger reserve under Project Tiger in 1994: Chundawat (2018).

Page 26, a respectable population of about thirty-one tigers: An estimated thirty-one tigers inhabited the Panna Tiger Reserve in 2001. See p. 211, Fig. 2 of Sen et al. (2009).

Page 26, two breeding-age females disappeared: Chundawat (2018), pp. 272–273.

Page 27, In 2003, a large male was found dead in a well: Chundawat (2018), pp. 274–275.

Page 27, "Losing them was like losing a part of one's family": Chundawat (2018), p. 5.

Page 27, coexist relatively peacefully with tigers: Kolipaka (2018) notes that the forty-two villages surrounding Panna Tiger Reserve have little conflict with tigers. For example, on p. 41, he writes, "Based on interviews, the researchers assessed that a large majority of the interviewed (79%) did not view tigers as a problem species. Only twenty one percent (21%) of the pastoralists viewed tigers as a threat, of which nineteen percent (19%) viewed them as threatening to large domestic animals."

Page 27, members of the Pardhi tribe were venturing deep into the park to hunt Panna's tigers: Sen et al. (2009).

Page 28, Pardhi hunters were generating income by supplying wildlife traffickers with a variety of animals: Sen et al. (2009), p. 16.

Page 28, "What we found, on our very first day there": Wright (2010).

Page 28, park managers then tried to silence him: Chundawat (2018), p. 87, describes how his research permits were withdrawn and curtailed following his efforts to notify the authorities about the decline of tigers in Panna Tiger Reserve.

Page 28, "Panna had lost nine breeding tigers out of eleven known to us": Chundawat (2018), p. 280.

Page 29, "estimated more than 30 tigers" still inhabited the reserve: Chundawat (2018), p. 281, reprinted an article written by Gopal, which asserted that the estimated population in Panna "was more than 30 tigers."

Page 29, staff counted "13 different adult tigresses within the territory of one tigress": Chundawat (2018), p. 283.

Page 29, a female wouldn't suffer a single rival in her territory: Chundawat (2018), p. 169, notes that female tigers in Panna defend exclusive territories.

Page 29, Researchers couldn't find a single tiger in Panna Tiger Reserve: Chundawat (2018), p. 295, describes how park staff found no definitive evidence of tigers in Panna after weeks of searching.

Page 29, India had an estimated 1800 tigers: Sharma et al. (2014) reviewed estimates of tiger abundance in India, including the estimate of 1800 in 1972 and 1411 in 2006.

Page 31, "not indulge in such stupid and senseless habits": Wright (2010), p. 96.

Page 33, three more animals were reintroduced over the next four years: Kolipaka et al. (2018) noted that Panna was kick-started with six successful tiger introductions.

Page 33, she had her first litter of at least three cubs: Kolipaka (2018) indicates that the female tiger named T2 had three cubs in October 2010. One of these cubs, a female named P213, had her own litter of cubs in the park, including P213(21), otherwise known as Panna Lal.

Page 34, doubling the population of adult tigers in one generation: T2's cubs were produced by at least two adults (T2 and at least one male); in order to replace themselves exactly, the breeding pair would need to produce two cubs to survive to adulthood. Having four survive doubled the population.

Page 34, thirty-one tigers were living in and around the park by 2018: Jhala et al. (2020), p. 40, Table 3.4.

Page 35, will each defend territories of around 7 square miles (20 square km): Sunquist (2010) notes, "The lower limit for female territory size appears to be about 15–20 km², a threshold probably set by social intolerance, not prey density." A territory of 20 square kilometers is equal to about 7 square miles.

Page 35, the park will be able to support only about thirty female tiger territories: At 280 square miles (540 square km), Panna Tiger Reserve could support about thirty female tigers. The total tiger population in the park could exceed thirty animals, however, because many cubs could reside within female territories, and some males may move between territories.

Page 35, six tiger reserves with an estimated total of 403 tigers as of 2018: Jhala et al. (2020) noted that 403 tigers lived in reserves (p. 40, Table 3.4) and 526 total tigers lived in Madhya Pradesh in 2018 (p. 30, Table 3.1).

Page 36, wildlife managers and conservationists are using sophisticated mapping software: Rathore et al. (2012) discussed how scientists are using sophisticated tools to identify potential corridors between parks in Madhya Pradesh.

Page 37, is connected to Panna by existing wildlife corridors: Dutta et al. (2016) examined the existing corridors between Madhya Pradesh's protected areas and noted that Panna and Noradehi parks have moderate connectivity.

Page 38, a growing list of networks designed to protect charismatic species: Examples of other park networks with corridors include the Yellowstone to Yukon conservation initiative in North America (https://y2y.net/), jaguar initiatives in Central and South America (for example, www.panthera.org/initiative/jaguar-corridor-initiative), and elephant initiatives in Africa (for example, https://africa geographic.com/stories/elephant-corridors-essential-for-the-species-and-environment/).

Page 39, already troubling signs that poachers are returning to Panna: According to an October 2020 story in *India Today* (www.indiatoday.in/magazine/up-front/story/20201012-alarm-bells-in-panna-1727783-2020-10-03), organized poaching rings were again working in the park, as several tigers had been killed by poachers.

Page 40, tiger populations have more than doubled to an estimated 2967 animals in 2018: Jhala et al. (2020), p. 30, Table 3.1, estimated India's tiger population at 2967 in 2018.

CHAPTER 2

Page 44, started their lives in this same gravel four years earlier: According to Professor Thomas Quinn at the University of Washington, most spawners in Hansen Creek have spent one year in Lake Aleknagik and just over two years in the ocean. When combined with the months they spent in the gravel while developing, their total lifespan is four years. Although a small percentage of Hansen Creek salmon return at different ages, in this chapter, I focus on the dominant four-year cycle for estimates of the success of reproduction.

Page 44, on a diet of zooplankton and larger foods such as squid and fish: According to Kaeriyama et al. (2000), sockeye eat a variety of food in the ocean, including larger prey such as fish and squid.

Page 46, Those daily counts comprised about as many salmon as returned in a typical year: From 1962 to 2013, the average return, based on University of Washington sampling methods, was 5432 sockeye. According to Thomas Quinn, the method used to count sockeye in the creek likely produced an underestimate, because UW used an index count in most years, which was the number of dead and live salmon in the stream on an index date (usually 6 or 7 August). Any salmon that returned later than the index date were not included in the total. In this chapter, I report only index counts for Hansen Creek.

Page 46, started measuring sockeye returns to this creek in the 1960s: Data on Hansen Creek salmon returns sent to me by Thomas Quinn include measurements going back to 1962.

Page 46, in 1966, with about 19,000 fish: UW researchers counted 18,696 sockeye in Hansen Creek that year.

Page 46, more than 55,000 sockeye salmon—ten times the average and three times the previous record: UW researchers counted 55,663 sockeye in Hansen Creek as of early August 2014, making this the index count. A total of 60,709 fish returned based on counts that extended later into August.

Page 47, in 1893, when it processed almost a million salmon: Professor Ray Hilborn of the UW's School of Aquatic and Fishery Sciences shared their sockeye salmon catch database going back to the initiation of the commercial fishery in Bristol Bay. They record 1893 as the first year with a commercial catch, which totaled 940,000 fish.

Page 47, catching ten million fish a year by 1901: The UW database records 10,220,577 sockeye caught in 1901.

Page 47, averaged more than fifteen million salmon per year: Based on the UW database, average harvest of sockeye in Bristol Bay from 1901 to 2015 was 15,596,380 fish.

Page 47, the captains of about 1300 boats descend upon Bristol Bay: Hilborn noted that Alaska Fish and Game issues about 1800 permits, but some boats fish on two permits so that they can catch more fish. He estimated that about 1300 vessels usually fish for salmon each year.

Page 48, the population size naturally oscillates around an average value: Note that the carrying capacity of any population is an estimate, and the actual population size can vary considerably over time. Furthermore, it's possible for the carrying capacity of a population to change over time as a result of shifts in the numbers of other species or changes to the physical environment.

Page 48, During a baby boom, large numbers of salmon survive to adulthood: Biologists typically refer to this kind of response as *density-dependent population growth*, where the rate of population growth literally depends on the density of the population (number of fish per defined area). In density-dependent population growth, small populations have higher growth rates resulting from mechanisms such as reduced competition or predation (predators move on to different, more abundant, kinds of prey). As the population grows, successful reproduction declines as a result of mechanisms such as increased competition or predation.

Page 48, a typical spawning fish today contributes not just one offspring to replace itself in the next generation, but three: Recruits per spawning fish can be very simply estimated by dividing the average total return (harvest plus spawning population) by the average spawning population. For the period between 1962 and 2019, these values are 18,945 total return divided by 5432 spawners, which equals 2.92.

Page 49, to sustainably harvest more than two billion salmon: The two-billionth sockeye was probably caught in the Bristol Bay Fishery in 2019, according to press coverage at the time. For example, a 2019 article from the *Bristol Bay Times* states, "This year, during the fishery's second-largest harvest on record, Bristol Bay commercial fishermen hit another historic number: the 2 billionth sockeye salmon caught by commercial fishermen since record-keeping began in the late 1800s." The full article is available at www.thebristolbaytimes.com/ article/1941bristol_bay_fishermen_catch_2_billionth.

Page 50, Sockeye choose mates in part based on their physical appearance: See Quinn et al. (2001) for a description of how body size is related to sexual selection and predation.

Page 51, with each growing and declining on its own schedule: Schindler et al. (2010) showed that the diversity of different salmon populations in Bristol Bay collectively dampened variation in salmon numbers over time.

Page 51, since 1980, the annual catch of sockeye has dropped below ten million only once: According to UW data, the fishery has landed fish every year since 1883, with the lowest annual catch occurring in 1973, when only 723,000 sockeye were landed. Since 1980, every year's catch, except 1998, has been greater than ten million sockeye.

Page 52, in 2020, 18.7 million sockeye eluded the fishery to spawn in Bristol Bay: The Alaska Department of Fish and Game estimated that 18.7 million sockeye escaped in 2020.

Page 54, rivers and lakes on Kodiak Island in the Gulf of Alaska: Beacham et al. (2006) identified lakes on Kodiak Island in the Gulf of Alaska and the Columbia River as likely refuge areas for sockeye salmon during the last ice age.

Page 55, many accounts of changes to salmon runs: The evidence for increased salmon runs in the Arctic is mostly anecdotal based on interviews of residents. Carothers et al. (2019) noted that some residents have observed that salmon are becoming more abundant, while others have not observed these changes.

CHAPTER 3

Page 57, The pygmy-possum ducked into a crevice in the rocks: Although this account of a pygmy-possum and an owl has been invented, it is based on what scientists know about the current biology of mountain pygmy-possums and their ancestral forest homes (for an overview, see Broome et al. 2012, Archer et al. 2019).

For example, after finding fossil bones in the remains of a cave (Broom 1896), scientists know that mountain pygmy-possums lived in this area thousands of years ago. According to the experts I interviewed, these fossil remains were likely included in pellets coughed up by an owl during the Pleistocene. Masked owls are known to roost in caves and deposit pellets on cave floors in southeastern Australia. Scientists have also roughly estimated the age of these fossils (late Pleistocene) and reconstructed how the environment was changing at the time. The fossils from the cave may have been from organisms that lived during an interglacial period, when the glaciers in the Northern Hemisphere retreated and the world warmed. During the Pleistocene, mountain pygmy-possums may have lived in wet forests such as the one described in this account, as other fossils found in this and other caves indicate that a variety of forest-dwelling species inhabited the area (Broome et al. 2012). Mountain pygmy-possums reproduce about once per year, usually giving birth to four young (Broome et al. 2012). The experts that I interviewed told me that the animals can carry objects, such as bedding for their nests, using their prehensile tails.

Page 59, a previously unknown species of small marsupial, which he named *Burramys parvus* **(small rock mouse):** Broom described *B. parvus* from jaw fossils he found in a cave. Broome et al. (2012) noted that the animal's scientific name means small rock mouse.

Page 59, but was instead a species of possum: Ride (1956) re-examined fossils of *Burramys parvus* and determined that it was not a close relative of rat-kangaroos, but was more likely a kind of possum.

Page 59, the possum and opossum lines diverged roughly eighty million years ago: Meredith et al. (2008) estimated that the American opossums split from the Australian marsupials about eighty million years ago.

Page 59, before the evolutionary split between dogs and cats: The estimate for the divergence of the domestic dog and cat (from www.timetree.org) is fifty-four million years ago.

Page 60, was quite different from any of the living pygmy-possums: Ride argued that *Burramys parvus* was related to other pygmy-possums, but it was different enough to warrant its own genus.

Page 61, their unexpected hutmate, which they named George: The possum's name was recorded in notes from Robert Warnecke, who transcribed some entries made by Alan Gilchrist in the University Ski Club log in October 1967. Gilchrist noted that, "Ken [Shortman] tried to identify 'George' from books" in Melbourne.

Page 62, (*Cercartetus nanus*), a common species in Victoria: Warneke told me that this species was widely distributed, relatively common, and easy to maintain in captivity.

Page 62, it "was quite illegal for me to do that": Regarding the gift of the eastern pygmy-possum to Shortman, Warneke told me that, "considering the unique circumstances[,] I willingly accepted responsibility for my decision."

Page 62, put the mysterious pygmy-possum into a locked cage: Warneke told me that when he received the pygmy-possum from Shortman, he initially focused on "establishing its dietary preferences before investigating its taxonomic affinities."

Page 62, the likes of which was known only from the fossil record: Warneke described the tooth that Wakefield saw in the captive pygmy-possum as "of a relative size and singular form previously known only from the fossils and sub-fossils of *Burramys parvis*."

Page 63, Two populations were also found farther away: Broome (2001) provides an overview of the living populations of mountain pygmy-possums that were found.

Page 64, they lived in lowland rainforests for millions of years: Archer et al. (2019) described the environment in which the ancestors of mountain pygmy-possums once lived.

Page 65, a few pygmy-possums managed to develop an entirely new lifestyle: The ability of mountain pygmy-possums to adapt to life in the snowy mountains was probably the result of two different elements of the rescue effect: phenotypic rescue and evolutionary rescue.

Page 65, one of its most important food sources is the Bogong moth: Broome et al. (2012).

Page 65, tasting "sweet and walnut-like": The descriptions of Bogong moths as "hamburgers on wings" that taste "sweet and walnut-like," come from a Zoos Victoria webinar on mountain pygmy-possums, at www.youtube.com/watch?v=fU_dhmH4idc.

Page 65, they have long been a food source for Aboriginal Australians: Stephenson et al. (2020) described long-term use of Bogong moths by Aboriginal Australians.

Page 66, "appearing to TV cameras like shimmering stars": Quote from *The New York Times*, www.nytimes.com/2000/09/24/olympics/bogong-moths-set-upon-olympic-park.html.

Page 66, meteorologists mistook it for a rain cloud: The anecdote that Bogong moths were mistaken for a raincloud, along with the quote about Yvonne Kenny, are from www.abc.net.au/news/2020-09-18/moths-panic-at-sydney-olympics-closing-ceremony/12674484.

Page 66, they complete their lifecycles at low elevations: Green et al. (2020) summarized the Bogong moth's lifecycle.

Page 66, the Bogong moth migration sets the calendar for their whole year: Broome (2001) described the importance of Bogong moths to mountain pygmy-possums.

Page 67, if they get too cold or they have insufficient fat reserves, they can die: Broome (2001) notes that mountain pygmy-possums need to build enough fat reserves and store enough food to survive winter hibernation.

Page 67, Mount Kosciuszko populations dropped 44 percent between 1997 and 2009: Broome et al. (2018) described the changes in the Mount Buller and Mount Kosciuszko mountain pygmy-possum populations.

Page 68, feral cats and foxes wait near the road to intercept pygmy-possums: Marissa Parrott of Zoos Victoria described the tendency of feral cats and foxes to hunt along the road to catch possums.

Page 69, climate warming is causing pygmy-possums to freeze to death: Although possums may freeze to death in the winter, this is probably mediated through starvation. The possums rely on stored food and body fat to help them generate extra body heat in winter. If they run out of either, they're likely to freeze as they starve to death.

Page 69, a result of changes in snowfall: Broome (2001) documented how reductions in winter snow can make pygmy-possums more susceptible to starving and freezing.

Page 69, which can drop to a frigid −4° F (−20° C): Linda Broome told me, "But without snow cover, the temperatures in the highest elevations in Australia, for example at Charlotte Pass, can get down to minus 20 degrees C."

Page 69, drought-stricken plants were unable to supply Bogong caterpillars with enough food: Green et al. (2020).

Page 71, the extinction resulted in part from rising sea levels: Gynther et al. (2016) confirmed that the Bramble Cay melomys was extinct and that sea-level rise was probably partly to blame.

Page 71, 94 percent of the island's vegetation had been killed by seawater: Gynther et al. (2016), p. 13, Table 1.

Page 71, seawater inundation and its affect on the island's vegetation had been documented: Gynther et al. (2016).

Page 71, a plan to rescue the animals in 2008: Woinarski (2016), for example, references the management plan for the Bramble Cay melomys, which was never implemented.

Page 71, the Australian government released a species recovery plan: The Australian government described a suite of actions that could help the mountain pygmy-possum recover (Department of Environment, Land, Water and Planning 2016).

Page 73, They provided "Bogong bikkies," tiny energy bars: Marissa Parrott told me that Zoos Victoria did extensive testing of Bogong bikkies with captive pygmy-possums to ensure that they would be available in the event of just such an emergency.

Page 73, an organism adjusts its phenotype to stay alive: The process of an organism adjusting to new conditions in the wild is usually referred to as *acclimatization* by scientists. I use the term *phenotypic rescue* to refer to a special case of acclimatization, where changes to an organism's phenotype enable it to survive under new conditions. This language also highlights the relationship between phenotypic rescue and the other ways that the rescue effect operates.

Page 74, "extinction of this high-elevation specialist species in the wild is inevitable": Archer et al. (2019).

Page 77, The hungry brushtails are infamous: Glen et al. (2012) described the adverse effects of brushtail possums in New Zealand.

Page 78, today's alpine possums are still likely to have many of the traits: Broome et al. (2012) and Archer et al. (2019) argue that mountain pygmy-possums may still have traits that will enable them to survive at low elevations.

Page 79, introduced species have caused significant harm to, and even extinction of, native species, especially in confined spaces: Thomas (2017) offers a good discussion of the effects of introduced species.

Page 79, "foreign species hardly ever cause native species to become extinct from entire continents": Thomas (2017), p. 110.

Page 80, the possums must rely primarily on a combination of reproductive rescue and phenotypic rescue: Mountain pygmy-possums could also theoretically evolve new traits to deal with a changing environment—that is, evolutionary rescue. However, there is no indication that this is happening.

CHAPTER 4

Page 84, "This disease was first noticed in the New York Zoological Park": Merkel (1906).

Page 85, dousing trees in antifungal chemicals to kill the blight: Merkel (1906) described futile efforts to try to stop the spread of the blight.

Page 85, the chestnut blight began spreading like wildfire: Powell et al. (2019) provides an overview of the spread of chestnut blight.

Page 85, The blight reached Connecticut, Massachusetts, and Washington, D.C., by 1908: Freinkel (2007), p. 41, describes the spread of chestnut blight to nearby states.

Page 85, "If this disease continues as it has begun": Murrill (1908).

Page 85, By 1912, the chestnut blight was present in at least ten states: Freinkel (2007), pp. 48–49.

Page 85, "The complete loss of the present commercial stand of chestnut in Pennsylvania": This quote, cited in Freinkel (2007), is from Sargent's "Letter of Transmittal," in the *Final Report of the Pennsylvania Chestnut Tree Blight Commission*, January 1 to December 15, 1913.

Page 86, "When from a mountain top one looks over thousands of acres of vigorous chestnut": Gravatt (1925).

Page 86, an estimated three to four billion chestnut trees were dead: This statistic has been referenced by many, including Brewer (2017), and is frequently cited by many of the experts that I interviewed. However, I have been unable to find the original source.

Page 87, the "perfect tree," with "a utilitarian versatility no other tree could match": Freinkel (2007), p. 25.

Page 87, "between 1907 and 1910, chestnut wood contributed more than ten million dollars annually": The quote is from Freinkel (2007), p. 26. I estimated the value of $10 million in 1910 in 2020 dollars using an inflation calculator (https://westegg.com/inflation/).

Page 87, "a good substitute for bread": Henry David Thoreau is quoted in Powell et al. (2019).

Page 87, they provided a rare opportunity as a cash crop to earn money: Freinkel (2007), pp. 21–24, noted the importance of the American chestnut as a cash crop for rural families.

Page 87, city dwellers would make excursions to the countryside to go "nutting": Freinkel (2007), p. 39.

Page 87, Native Americans saw similar food value in chestnuts: Freinkel (2007), pp. 16–17, noted the value of chestnuts to Native Americans, who maintained chestnut orchards.

Page 88, Several species of moth relied exclusively on the American chestnut: Opler (1979) described some of the insects that relied on American chestnut; of the sixty species of moths known to feed on chestnuts, seven apparently fed exclusively on the tree. Some of these species may now be extinct.

Page 89, a great forest extended across the northern reaches: Deng et al. (2015) offers an overview of the theory that many Northern Hemisphere trees were once part of a continuous northern forest.

Page 89, The ancestors of modern chestnut trees were eventually divided into separate populations: Lang et al. (2007) concluded that the division between chestnuts in Asia, Europe, and North America occurred in the middle to late Eocene, which is roughly thirty-five million to forty-seven million years ago.

Page 89, how many species were ultimately produced though this process: Lang et al. (2006) mentions some of the controversy regarding the number of chestnut species and describes the phylogenetic relationships between the different groups.

Page 90, Asian chestnut species encountered a fungal pathogen, *Cryphonectria parasitica*: Powell et al. (2019) notes that the chestnut blight likely evolved in Asia before being introduced to North America.

Page 91, each side incrementally evolving to deal with the other more successfully: Anagnostakis (1987) summarized how hosts and parasites can co-evolve in ways that result in epidemics when the parasite encounters a naïve species.

Page 91, the Asian trees can usually contain the damage caused by the fungus but cannot defeat it entirely: Asian species are typically able to survive in the presence of chestnut blight. For example, Steiner et al. (2017) showed that, while susceptible to the disease, Chinese chestnut is able to defend itself and survive.

Page 91, people in the United States had been importing exotic chestnut trees: Powell et al. (2019) noted that, "chestnuts trees, primarily Chinese chestnut, have been widely distributed for over 100 years and many are planted in orchards throughout the United States for nut production," which is presumably what led to the original introduction of the chestnut blight in North America.

Page 91, it produces small, orange to yellow bumps that erupt through the tree's bark: Murrill (1908) described the progression of the chestnut blight disease.

Page 91, with the ability to be carried by the wind: During an interview, William Powell told me that the sexually produced spores are better at wind dispersal than the asexually produced spores.

Page 94, "we're now sitting at about 430 million" of these resprouting root systems: Fitzsimmons's estimate is based on the research of Dalgleish et al. (2016).

Page 94, These infected sprouts then become new sources for fungal spores: Prospero et al. (2006) observed that even dead chestnut trees can be sources of spores that can infect living trees.

Page 94, the fungus has already spread to the trees in Wisconsin: A stand of American chestnut thrived in Wisconsin for many years before eventually being negatively impacted by the fungus (Double et al. 2018).

Page 95, the organism in question must experience differential mortality: An increase in the death rate is not technically necessary for evolution to occur if differential reproduction is present. For example, suppose individuals that are more fit reproduce, while reproduction is hindered among the less fit. The next generation will be disproportionally represented by those individuals that are more fit.

Page 97, breeding has yet to produce trees with enough resistance: Barakat et al. (2012) note that selective breeding by the American Chestnut Cooperators' Foundation has yielded only a limited amount of disease resistance.

Page 97, breeders are searching for ways to find new genes from other species:
Two related concepts in genetics are relevant here. A gene is a unit of DNA that
codes for some particular end product, often a protein. The same gene can be written slightly differently, producing a different product, and these variations in the
same gene are called *alleles*. Plants and animals have two copies of most of their
genes, one from each parent, which can be different from each other (that is, they
may have two different alleles). To fight the blight, American chestnut trees may
need entirely new genes; it's also possible that they just need some new alleles for
existing genes. For simplicity, I refer to the need for new genetic combinations as
"genes," rather than drawing a distinction between genes and alleles.

**Page 97, breeders have hybridized thousands of surviving American chestnuts
with Asian chestnut species:** Powell et al. (2019) noted that Yale Professor Arthur H.
Graves was one of the first to do this, in 1921. A parallel effort was undertaken by the
USDA in 1922.

Page 98, However, the hybrids never reached a desired trifecta of traits: Powell et
al. (2019), p. 4, notes that none of the early efforts to hybridize chestnuts "succeeded
in producing a fast growing, timber-type tree with good blight resistance."

**Page 98, hybrids probably needed to carry two copies—one from each parent—
of many of the resistance genes:** In sexual reproduction, offspring inherit different copies of the same gene from each parent. In some cases, to work properly,
a gene needs both copies to be the same. In first-generation hybrids, both copies
cannot come from only one of the parent species. Because blight resistance seems
to require that both gene copies be inherited from an Asian species, full blight resistance could be produced only by breeding multiple generations of hybrids.

Page 98, while retaining only the stripes from tigers: Backcrossing of striped
lion-like cats might work especially well if tiger stripes are the result of several genes
that work together to produce the striped phenotype.

Page 99, breeders have been trying to backcross a blight-resistant chestnut tree:
Steiner et al. (2017) describes the process of inoculating trees to test their levels of
blight resistance.

**Page 100, they thought the disease-resistance of Chinese chestnuts might be
controlled by just two or three genes:** Steiner et al. (2017) summarized Dr. Charles
Burnham's plan for breeding a resistant hybrid, which assumed that disease resistance was controlled by two or three genes.

Page 100, They now know that there are at least twelve resistance genes: Dr. Jared Westbrook, director of science at The American Chestnut Foundation, told me that chestnut blight resistance in Asian chestnut trees is genetically complex and is controlled by genes on all twelve of the chestnut's chromosomes.

Page 102, laboratory genetic engineering is merely an example of biomimicry: This use of the term *biomimicry* is distinct from the process of mimicry in nature, whereby one species evolves to resemble another. In Batesian mimicry, for example, palatable species evolve to resemble others that are less palatable to improve their chances of survival.

Page 102, scientists discovered that this phenomenon has actually occurred in sweet potatoes: Kyndt et al. (2015) showed that the genome of the cultivated sweet potato contains *Agrobacterium* genes.

Page 103, such as aphids that have genes from a fungus and ferns that have genes from hornwort plants: Moran and Jarvik (2010) showed that some aphids have genes from a fungus, and Li et al. (2014) showed that some ferns have genes from hornwort plants.

Page 103, scientists inserted a gene from a firefly into a tobacco plant: Ow et al. (1986).

Page 104, people didn't like the idea of an American chestnut with "frog genes": Powell et al. (2019), p. 6, described the perception problem with using a frog-inspired gene to create a disease-resistant American chestnut tree: "Though the peptides used were synthetic in origin, and not cloned directly from a frog or any other organism, they became known as the 'frog genes' with all the problematic public perceptions that this engendered."

Page 104, he had his "eureka moment . . . while reading a book of abstracts": Powell et al. (2019).

Page 105, they have all 30,000-plus American chestnut genes, plus the single gene from wheat: In addition to the wheat gene, Powell told me, "Our current version (Darling 58) has the oxalate oxidase gene which confers blight tolerance and a selectable marker gene NPT2. The NPT2 gene is only used to select for the cells that have taken in the genes during the transformation and is one of the oldest marker genes."

Page 106, starting to increase the gene pool of blight-tolerant American chestnuts: Westbrook et al. (2020) describes the plan for increasing the diversity of transgenic chestnut trees.

Page 106, may begin in earnest during the next decade: In fact, the backcrossed hybrid trees are already being planted in scientific trials.

Page 107, Is it acceptable to propagate them?: Newhouse et al. (2020) wrote and submitted a petition to the Animal and Plant Health Inspection Service (APHIS) to allow unregulated planting of a transgenic American chestnut tree.

Page 107, the Food and Agricultural Organization of the United Nations has listed eleven separate potential risks of transgenic organisms: The potential risks of GMOs compiled by the FAO are available at www.fao.org/english/newsroom/focus/2003/gmo8.htm.

Page 108, "In the absence of credible independent long-term feeding studies, the safety of GMOs is unknown": From the Non GMO Project's website, at www.nongmoproject.org/gmo-facts/.

Page 112, genetic engineering has been proposed to help endangered black-footed ferrets: Novak et al. (2016) called for using genetic engineering tools to rescue the black-footed ferret from extinction.

CHAPTER 5

Page 115, crammed with roughly 25 percent of all ocean species: Knowlton et al. (2010) estimated that a quarter of all marine species can be found in coral reefs.

Page 120, Spawning at night helps corals avoid predators: Some predation by fish does happen at night, but many more fish are present during the day, and they use daylight to find small pieces of food—such as eggs and embryos—in the water column.

Page 121, One study, for example, estimated the value of the island's natural environment, which is dominated by the ocean, at $1.3 billion: Taylor et al. (2011).

Page 121, half a billion people rely on coral reefs for food and economic opportunities: Hoegh-Guldberg et al. (2019) estimated that half a billion people "have developed a high degree of dependency on ecosystem goods and services provided by coral reefs."

Page 122, live corals covered about 50 percent of Caribbean reefs: Gardner et al. (2003) showed that coral covered about 50 percent of the Caribbean reefs in the 1970s.

Page 122, a typical Roatán reef included about 30 percent live coral cover in 2018: This estimate comes from the Healthy Reefs Initiative's Mesoamerican Reef Data Explorer, at https://oref.maps.arcgis.com/apps/MapSeries/index .html?appid=9556c100e1d9424fa9b3c1748454e297.

Page 123, One group of fishes, the grazers, benefit corals by mowing down seaweeds: In addition to fish, many other species of invertebrates, such as sea urchins, act as grazers on coral reefs.

Page 123, often killing entire colonies within weeks or months: Muller et al. (2020) includes an overview of what is known about stony coral tissue loss disease.

Page 123, Locals first noticed it in Roatán in September 2020: Jenny Myton told me that stony coral tissue loss disease was first observed in Roatán in September 2020.

Page 123, as much as 90 percent of our extra heat has been absorbed by the ocean: Zanna et al. (2019).

Page 125, Prior to 1980, coral bleaching was observed as a rare, localized event: According to Hughes et al. (2018), coral bleaching was a rare, localized event prior to the 1980s; today it affects nearly every reef on the planet.

Page 125, Roatán's reefs have bleached at least nine times since 1995: The Healthy Reefs Initiative cites coral bleaching events occurring in Roatán in 1995, 1998, 2005, 2010, 2015, and 2016, in the site's Coral Bleaching tab at https://oref.maps.arcgis. com/apps/MapSeries/index.html?appid=9556c100e1d9424fa9b3c1748454e297. In addition, I have personally observed bleaching in Roatán in 2017 and 2019, and I learned through contacts in Roatán that bleaching occurred again in 2020.

Page 126, between 2006 and 2016, the abundance of grazing fish doubled in the reserve: Although fish numbers have historically increased in West End, the Healthy Reefs Initiative's most recent data from 2018 indicates that fish populations declined relative to 2016. It's not clear whether this resulted from increased poaching, sampling error, or something else.

Page 130, genetically predisposed to deal with high temperatures, pollution, or other stressful conditions: Some corals are better at withstanding stress than others. For example, Palumbi et al. (2014) showed that different coral colonies from the same species can differ markedly in their tolerance of high temperatures.

Page 130, the temperature threshold for coral bleaching has increased by about 0.5°C: Sully et al. (2019).

Page 131, rescue would be strong enough to help corals keep pace with climate warming in the future: Demographic, reproductive, and evolutionary rescue were all explicit parts of the model. Phenotypic rescue was included indirectly, in that reefs that were well managed locally included corals that were more likely to survive by outcompeting seaweeds for space.

Page 131, We used mathematical models: The mathematical model of how corals may adapt to global warming is described in Walsworth et al. (2019).

Page 133, the cost of restoring reefs through coral gardening at $351,661 per hectare: Bayraktarov et al. (2019).

Page 133, the world has 96,500 square miles (250,000 square km) of coral reef habitat: Burke et al. (2011).

Page 133, Restoring all of that would result in a jaw-dropping price tag of $9 trillion: The estimated 250,000 square km of reef habitat is equivalent to 25,000,000 hectares. By multiplying 25,000,000 hectares of habitat by $351,661 per hectare, the net result is $8.79 trillion.

Page 134, some super corals that occur naturally on reefs: Palumbi et al. (2014), for example, have identified heat-tolerant super corals in American Samoa.

Page 134, breeders can artificially select for traits such as thermal tolerance: Van Oppen et al. (2015) provides an overview of techniques for assisted evolution on coral reefs.

Page 134, a recent modeling study to estimate the amount of effort needed to boost the evolution of corals: DeFilippo et al. (2022) examined the effects of restoration and assisted evolution on the fate of corals.

Page 139, Orange cup corals from the Indo-Pacific were first observed: Riul et al. (2013) provides an overview of the colonization of orange cup corals in the Caribbean and discusses their effects on native species.

CHAPTER 6

Page 145, Tijs Goldschmidt had an epiphany: Goldschmidt (1996), p. 201, provides an account of his epiphany about the Nile perch in Lake Victoria.

Page 146, he experienced "a feeling of disorientation, of bewilderment": Goldschmidt (1996), p. 201.

Page 146, was colliding with the events of the here and now: Goldschmidt (1996), p. 202, describes that his focus at the time was on the evolution of the fish species in Lake Victoria, rather than on the fact that fish populations were starting to decline.

Page 146, Perhaps the most famous rift zone in the world is in eastern Africa: Chorowicz (2005) describes the formation of the East African Rift Valley.

Page 147, forming what is now Lake Victoria: Johnson et al. (2000) describes the formation of Lake Victoria.

Page 147, Lake Victoria has oscillated between being a large lake when rains were plentiful: Johnson et al. (2000) describes the cycles of Lake Victoria drying out and filling.

Page 147, two species of haplochromine cichlid fish had a chance encounter: Meier et al. (2017a) suggests that ancestors of the Lake Victoria haplochromine cichlids were ancient hybrids.

Page 147, two related species of cichlids came into contact after millions of years: Meier et al. (2017a) estimates that the two cichlid lineages split between 1.6 million to 5.8 million years ago.

Page 148, the surface of Lake Victoria grew larger than the surface area of Switzerland: Barel et al. (1985).

Page 149, Some evolved into zooplanktivores: Goldschmidt (1996), pp. 27–36, describes many feeding specializations developed by different species of Lake Victoria haplochromines.

Page 149, All of Lake Victoria's haplochromine cichlids are maternal mouthbrooders: Goldschmidt (1996), p. 33.

Page 150, covering her snout with its huge, rubbery mouth: In an interview with Les Kaufman, he confirmed that some paedophages engulf the snouts of mouthbrooding females to feed on their young.

Page 150, at least 150 unique cichlid species occupied the lake: Witte et al. (1976) cited an estimate of 150–170 species from a 1974 study by P. H. Greenwood: "The cichlid fishes of Lake Victoria, East Africa: The biology and evolution of a species flock," published in the *Bulletin of the British Museum of Natural History (Zoology)*, Supplement 6.

Page 150, at least 200 species were present: Witte et al. (1976)

Page 150, the estimate grew to 300 or even 400 species: Witte et al. (1992) and Kaufman (1992).

Page 150, at least 500 species must have evolved: Seehausen et al. (1997).

Page 152, females tend to be very particular about the appearance of their would-be suitors: Seehausen et al. (1997) describes how female mate choice helps keep populations of closely related cichlids from interbreeding, even when they live in the same place.

Page 152, *Pundamilia* haplochromine species that are living on the same rocky reef have different preferences: *Pundamilia* is a genus of haplochromine cichlids that includes multiple species. One of the species with blue males is *P. pundamilia*, while one of the species with red males is *P. nyererei*. Meier et al. (2017b) concluded that these two species likely evolved side-by-side, as evolution favored blue males in shallow water and red males in deeper water. Remarkably, Meier et al. (2017b) concluded that similar evolutionary patterns may have happened many times, so that different populations of *Pundamilia* species separately evolved their own blue and red male forms.

Page 152, this has given rise to groups of closely related species: Marques et al. (2019) described some of the groups of species that have diversified into many closely related forms.

Page 152, the speed with which this has occurred: McGee et al. (2020) concluded that the Lake Victoria haplochromine cichlids have undergone exceptionally fast diversification and speciation.

Page 153, "the most exuberant" example of this kind of evolution in the world: Kaufman (1992).

Page 153, it was endowed with essentially two complete sets of genetic variations: Meier et al. (2017a) concluded that the rapid evolution of Lake Victoria cichlids was related to their ancestral hybridization.

Page 154, it gobbled up the haplochromines so fast that it overwhelmed the rescue effect: Marshall (2018) concluded that Nile perch predation was the best explanation for the decline of the haplochromine cichlids in Lake Victoria.

Page 154, The largest individuals can grow to almost 6 feet (2 m) long: According to FishBase (www.fishbase.se/Summary/SpeciesSummary.php?ID=347&AT=-nile+perch), the maximum length of a Nile perch is 2 meters and it can weigh as much as 200 kilograms.

Page 154, Nile perch were historically confined to the lower parts of the river: Several experts noted that Nile perch, or a close relative, once lived in Lake Victoria, but the fish disappeared from the lake thousands of years ago, perhaps during one of the lake-drying events.

Page 154, "whether fishery productivity might be increased by introducing non-indigenous species to Lake Victoria": Pringle (2005), p. 781.

Page 155, they were "'trash fish' of very little value": Pringle (2005) cited this from an article written by A. M. Anderson, "Further observations concerning the proposed introduction of Nile perch into Lake Victoria," published in the *East African Agricultural and Forestry Journal* in 1961.

Page 155, "based on ignorance of several fundamental biological concepts": Fryer (1960), p. 270.

Page 155, "such an introduction is not only undesirable, but would jeopardize the existing commercial fishery": Fryer (1960), p. 270.

Page 155, this wasn't documented until almost a quarter century later: Pringle (2005), p. 784, noted that the first clear documentation that he could find of the 1954 Nile perch introduction came in a letter to the *East African Standard*, published in 1978.

Page 156, totaling a few hundred individual fish: Pringle (2005), p. 784, documented the known introductions of Nile perch to Lake Victoria in the 1960s.

Page 156, some reports indicated that they were already breeding: Pringle (2005), p. 783, cited the Uganda Game and Fisheries Department's conclusion that the Nile fish were breeding in Lake Victoria as early as 1962.

Page 156, the number of Nile perch had increased 10,000-fold since its introduction: Marshall (2018), p. 1544.

Page 156, significant numbers of Nile perch were recorded in Ugandan fisheries: According to Marshall (2018), significant numbers of Nile perch were recorded in Ugandan fisheries by the mid-1970s.

Page 156, Kenyan fisheries reported similar increases by around 1978: Figure 3 in Kaufman shows that Nile perch catches in Kenya started ramping up in 1978.

Page 156, it appeared that the haplochromine cichlids might be able to hold their own: Kaufman (1992) showed that haplochromines were still being harvested in large numbers into the late 1970s in Kenya. Marshall (2018) showed large numbers of haplochromine harvests extending until the early to mid-1980s in Tanzania.

Page 156, by 1985, haplochromines had declined to just 7 percent of the catch: Marshall (2018), p. 1545, Table 2, provides data regarding declining catches of haplochromines and increasing catches of Nile perch.

Page 157, "an enormous loss to evolutionary biology": Barel et al. (1985).

Page 157, fully 80 had disappeared by 1986—a 65 percent decline: Witte et al. (1992) documented changes in the species composition in the Mwanza Gulf after the rise of Nile perch populations.

Page 157, fisheries for the haplochromines and other species had probably already reduced their numbers: Witte et al. (2000), for example, noted that intensive fisheries for haplochromine cichlids existed in Lake Victoria at the time that Nile perch numbers increased in the 1980s.

Page 157, more pollution, which made the lake murkier: Seehausen et al. (1997) described the decline in water clarity in Lake Victoria.

Page 157, affected the availability of oxygen in the lake: Marshall (2018) noted that low oxygen levels were harming the Nile perch.

Page 157, fish had also been introduced to provide new opportunities for fishing: For example, Kaufman (1992) noted that several larger species of cichlids were introduced to Lake Victoria—not haplochromines, but a different group of cichlids commonly known as tilapia.

Page 157, changes in water clarity may have significantly affected the losses of distinct species: Seehausen et al. (1997).

Page 158, when water pollution "turns off the lights in the lake": Seehausen et al. (1997), p 1810.

Page 158, "the first mass extinction of vertebrates that scientists have ever had the opportunity to observe": Kaufman (1992), p. 86.

Page 158, "Never before in recorded history had the mass extinction of vertebrate organisms occurred on such a scale or within such a time frame": Goldschmidt (1996), p. 207.

Page 158, the rise of the Nile perch was accompanied by a tenfold increase in fish catches in Lake Victoria: The rise in fish catches from Lake Victoria that coincided with the rise of the Nile perch has been well documented. For example, Kaufman (1992) discusses the dramatic increase in catches in Kenya in the 1980s.

Page 159, an international export market that was worth about $250 million annually by the 2000s: Odongkara et al. (2010) showed that Nile perch landings were worth $241 million by 2003 and remained at this value or higher until 2008.

Page 159, created both winners and losers, economically: Both Kaufman (1992) and Barel et al. (1985) noted that some fishers gained while others lost because of the increases in Nile perch in Lake Victoria.

Page 160, fishing pressure has arguably become too high, resulting in overharvesting: Matsuishi et al. (2006) argued that Nile perch were arguably being overharvested by the 2000s.

Page 161, Dutch researchers originally called the "orangehead": Orangeheads were scientifically described in Witte and Witte-Maas (1987) and named *Haplochromis (Yssichromis) pyrrhocephalus*; the species name describes the flamelike (*pyr*) orange pigmentation on the heads (*cephalus*) of the males.

Page 161, orangeheads were usually about 2–3 inches (5–7 cm) long: The fish measured by Witte and Witte-Maas (1987), p. 38, ranged from 56 to 72 mm, which is roughly 2–3 inches. The description of their habitat preferences and diet appears on p. 40.

Page 161, in 1979, orangeheads were found to be one of the most abundant zooplanktivore species in the Mwanza Gulf: Witte et al. (2000), p. 235, Table 1, illustrates the abundance of orangeheads found in the 1970s in the Mwanza Gulf.

Page 161, orangehead populations had been reduced by over 99.99 percent since the late 1970s: Witte et al. (2000), Table 2, shows a drop in catch averages from 15,987 fish per hour with large bottom trawls in 1978, to 0.6 fish per hour in 1987.

Page 161, As a whole, the species group had declined by about 99.5 percent: Witte et al. (2000), Table 1, shows that zooplanktivore catches dropped from an average of 149.8 fish per trawl in 1979 to an average of 0.7 fish per trawl in 1987–88, which is a 99.5 percent decrease.

Page 161, orangehead populations had become nearly 18,000 times more abundant than during their low point in 1987: Witte et al. (2000), Table 2, shows an increase from average catches of 0.6 fish per hour with large bottom trawls in 1987, to 10,760 fish in 1997—a 17,933-fold increase.

Page 161, by 2006, they were more than 10 times more abundant than they had been in the late 1970s: Kishe-Machumu et al. (2015), Table 3, shows that the abundance per transect of orangeheads was 27 in 1979–80, 551 in 2006, and 230 in 2008.

Page 161, overharvesting of Nile perch in the 1990s removed most of the largest fish that had been preying on the orangeheads: Several researchers, including Witte et al. (2000) and Matsuishi et al. (2006), concluded that high harvest levels of Nile perch helped the haplochromines recover in the 1990s.

Page 162, the orangeheads rebounded through reproductive rescue: Kishe-Machumu et al. (2015) noted the return of species such as orangeheads after intensive fisheries increases for the Nile perch.

Page 162, the orangeheads of the 2000s occupied more territory: Kishe-Machumu et al. (2015) noted the orangeheads' expansion into new habitat.

Page 162, by the 1990s, they were eating larger fish, shrimp, and mollusks: Katunzi et al. (2003) documented the changes in orangehead diets.

Page 162, with a smaller head and a larger tail: Van Rijssel and Witte (2013) documented the changes in orangehead body shape and speculated about the causes of these changes.

Page 162, These changes to orangeheads occurred very quickly, within only a decade: According to van Rijssel and Witte (2013).

Page 162, the orangeheads' bodies began to revert back to their original shapes: According to van Rijssel and Witte (2013).

Page 163, the fish are hybridizing with one of their close relatives: Mzighani et al. (2010) concluded that orangeheads and another cichlid species, *Haplochromis laparogramma*, were hybridizing in the Mwanza Gulf.

Page 164, One species that specializes in eating mollusks in shallow water appears to be mixing with another species that tends to eat detritus: When I spoke with Jacco van Rijssel, he identified the two species that are hybridizing as *Platytaeniodus dengeni* and *Haplochromis antleter*.

Page 165, surviving cichlids in Lake Victoria lack some of the feeding specializations of their recent ancestors: Ole Seehausen told me that the loss of ecological specialization has only been observed "for species in the demersal [near the bottom] offshore zone of the lake."

Page 165, mollusk-eating and detritivore populations were thriving by the mid-2000s: Kishe-Machumu et al. (2015) observed that, as a group, the haplochromine populations had largely returned to their 1970s abundance in the Mwanza Gulf by the mid-2000s, and that representatives of nine of twelve functional groups had also returned (the missing functional groups were prawn-, scale-, and parasite-feeding specialists).

Page 166, and they have remained at this higher level: Njiru et al. (2018) compiled fisheries data for Lake Victoria, showing how haplochromine fisheries declined in the 1980s, but then returned larger than ever before in the early 2000s and remained so through 2015 (see Fig. 1B). Of course, fishery catch does not necessarily indicate higher abundance, especially if there have been increases in fishing efforts or better reporting since the 1970s. However, the higher catch is consistent with the resurgence of haplochromines noted in detailed studies in the Mwanza Gulf. For example, Kishe-Machumu et al. (2015), Table 2, observed that haplochromines in the Mwanza Gulf were similarly abundant in 1979–80, 2006, and 2008.

Page 166, these fish are primarily consumed and valued locally: One researcher told me that, although local fishers were capable of seeing differences between the species, especially males in breeding colors, they tended to lump the haplochromines into a single group they called *furu*.

Page 168, "has become a kind of ballad, warning of the dangers of meddling with nature": Kaufman (1992), p. 851.

Page 168, hundreds of unique species now extinct because of human meddling: Witte et al. (1992) estimated that 200 species had either gone extinct or were in danger of doing so by 1990.

Page 169, Chinook salmon populations that had been introduced to New Zealand were evolving to adapt to local conditions: Quinn et al. (2001) reviewed evidence that Chinook salmon introduced to New Zealand adjusted and evolved to match local conditions within thirty generations.

Page 169, black snakes in Australia have rapidly evolved to deal with an invasion of toxic cane toads: Phillips and Shine (2006) showed that black snakes evolved abilities to deal with toxic cane toads within twenty-three generations.

Page 169, the species creation rate is exceeding the extinction rate: Thomas (2015) reviewed evidence for the rise of new plant species in Europe.

Page 171, the species has formed at least nine different groups that may well be on their ways to becoming new species: Moser et al. (2018) documented the diversification of one haplochromine species into nine different forms (or morphs) within fifty years.

CHAPTER 7

Page 173, but this one was larger than anything they had ever witnessed: The account of the discovery of Chauvet Cave is detailed in Chauvet et al. (1996). The description of the bear bones and nests and their observation that "caves this large were totally unknown in the gorges," appear on p. 36.

Page 174, Prior to their extinction about 24,000 years ago: Terlato et al. (2019) estimated that cave bear bones found in Italy were from animals that lived about 24,200 to 23,500 years ago, leading them to conclude that these were the latest known individuals of the species.

Page 174, Within the sea of bones, the explorers noticed large depressions in the earth: Chauvet et al. (1996), p. 36

Page 174, two unmistakable lines of red ochre: Chauvet et al. (1996), p. 40.

Page 174, _"Ils sont venus!"_: Eliette Brunel Deschamps is quoted in a video about the cave discovery on the Chauvet Cave website, at https://archeologie.culture.fr/chauvet/en.

Page 175, Deeper in the cave, they found more drawings: Chauvet et al. (1996) includes many descriptions and photos of various panels in the cave. Not all of the artwork and artifacts in the cave were observed on their first trip inside. The team returned later on the same night for a second exploration and returned a week later to document the drawings more completely with photographs and video.

Page 175, "We thought we could feel their presence; we were disturbing them": Chauvet et al. (1996), p. 42.

Page 176, they were the first visitors—human or bear—in more than 20,000 years: According to the Chauvet Cave website (https://archeologie.culture.fr/chauvet/en/immediate-protection-measures), the cave entrance was blocked by a rockslide at least 20,000 years ago.

Page 176, making these among the oldest known drawings in the world: The cave artists likely visited the cave between about 30,000 and 36,500 years ago, according to the Chauvet Cave website at https://archeologie.culture.fr/chauvet/en/aurignacian-cave-art.

Page 177, climate change was a factor in many of the extinctions: MacPhee (2018) discusses how the climate was changing at the end of the Pleistocene.

Page 177, modern humans and closely related species such as Neanderthals—had something to do with the extinctions in Europe and Asia: MacPhee (2018) summarizes the arguments about whether the Pleistocene extinctions were caused by climate change, hunting by early humans, or a combination of the two.

Page 177, perhaps something similar happened as modern humans first reached the Americas and Australia: Scientists have found no evidence that hominins other than modern humans ever reached the Americas or Australia.

Page 178, Cave bears roamed throughout Europe and into Asia for hundreds of thousands of years: Barlow et al. (2021) showed that cave bears diverged from their polar and brown bear ancestors about 1.5 million years ago. The lineage that included *Ursus spelaeus* also included several other distinct populations that had been largely evolving on their own some 300,000 to 500,000 years ago, and they may have been diverging into different species.

Page 178, cave bears were mostly vegetarians: Pacher and Stuart (2009) summarize the scientific consensus that cave bears were largely vegetarian and would have required substantial food resources to support their large body size.

Page 178, males grew to 1100 pounds, or 500 kg: Christiansen (1999) estimated that adult cave bear females weighed about 225–250 kg (about 500–550 pounds), with males reaching 400–500 kg (about 900–1100 pounds).

Page 178, the animals' genetic diversity and population size were starting to decline about 50,000 years ago: Stiller et al. (2010) argue that cave bear population likely started to decline 25,000 years before the species went extinct.

Page 178, the bears may have become trapped between glaciers in the mountains and the overly dry steppe grasslands below: Döppes et al. (2019) concluded that, during the late Pleistocene, cave bears became trapped on mountainsides between dry grasslands below and icy conditions above, with most populations disappearing between 35,000 and 25,000 years ago.

Page 178, At least a few holdouts survived for another 1000 years: Terlato et al. (2019) discusses evidence that indicates that cave bears survived in Italy until 24,000 years ago.

Page 179, scientists have postulated that hominins were also involved: Terlato et al. (2019) concluded that hominins were probably involved in the extinction of cave bears.

Page 179, people who occupied caves like Chauvet deprived the bears of their much-needed winter shelter: Mondanaro et al. (2019) and Grayson and Delpech (2003).

Page 179, cut marks on cave bear bones, indicating that hominins at least occasionally butchered, and perhaps hunted, the animals: Terlato et al. (2019).

Page 180, the last few individuals dying on Wrangel Island in the Arctic Ocean about 4000 years ago: Vartanyan et al. (1995) estimated that woolly mammoths survived on Wrangel Island until around 2000 BC.

Page 181, tools that could be used to accomplish this are cloning and gene editing: Shapiro (2015) summarizes the science of de-extinction, including the two primary tools that could be used to accomplish this: cloning and gene editing.

Page 181, they had succeeded in cloning a mammal, a sheep later named Dolly: Wilmut et al. (1997) describes the successful cloning of lambs using nuclear transfer.

Page 181, young Dolly, shared the nuclear genetic code of her first mother, the donor ewe: Dolly got most of her DNA from her first mother. However, her mitochondrial DNA, which is stored outside of the nucleus, came from her second mother. It's also possible that Dolly had some differences from her first mother in the code of her nuclear DNA because of mutations that naturally occur as the body makes more cells.

Page 182, Cave bears evolutionarily split from brown bears about 1.5 million years ago: Barlow et al. (2021) estimated that cave bears split from brown bears and polar bears about 1.5 million years ago. However, there was probably some hybridization between brown bears and cave bears as recently as a million years ago.

Page 182, The closer the donor and surrogate are to the species to be cloned: Shapiro (2015), p. 147.

Page 182, gray wolves have been successfully cloned this way: Borges and Pereira (2019) reviewed the use of cloning in conservation and summarized efforts for wolves, Bactrian camels, and bucardos.

Page 183, They have since attempted to clone the animal: Folch et al. (2009).

Page 184, the ancient DNA is certain to be contaminated with the DNA from other organisms: Shapiro (2015), p. 63.

Page 186, they used 277 eggs, 29 embryos, and 13 surrogate mothers: Wilmut et al. (1997), p. 811, Table 1.

Page 186, efforts to clone a bucardo have been similarly inefficient: Folch et al. (2009) conducted two cloning experiments—the first starting with 285 embryos, and the second with 154 embryos—to produce a single "morphologically normal" newborn, which died within minutes of its Cesarean birth.

Page 186, each species poses its own unique challenges: Borges and Pereira (2019) discuss many challenges of efforts to clone different species.

Page 189, It is "ecological resurrection, and not species resurrection, that is the real value of de-extinction": Shapiro (2015), p. 131.

Page 189, Mammoths probably helped create and maintain their steppe ecosystem: Zimov (2005) describes the ecological role of the mammoth in maintaining a northern grassland.

Page 190, more than twice the carbon that is currently held in all of the world's rainforests combined: Zimov (2005), p. 798, noted that, "the soil of the mammoth ecosystem harbors about 500 gigatons of carbon, 2.5 times that of all rainforests combined."

Page 190, Zimov and his colleagues are rebuilding an ecosystem that resembles the steppe grassland ecosystem: Zimov (2005).

Page 191, The maximum population growth rate for African elephants is about 7 percent per year: Calef (1988).

Page 191, Zimov believes that about 1 million square kilometers: Zimov (2005), p. 798, notes that, "In northern Siberia, mainly in the Republic of Yakutia, plains that once were covered by tens of meters of mammoth steppe soils now occupy a million square kilometers."

Page 191, that steppe ecosystems once supported around one mammoth per square kilometer: Zimov et al. (2012), p. 211.

Page 192, existing grazers can work especially well to restore the steppe environment: Beer et al. (2020) provides evidence that grazing reduces soil temperature.

Page 192, they can construct reasonable-seeming justifications for pursuing this outcome: Kunda (1990) described how people are prone to arriving at their preferred conclusion by "constructing seemingly reasonable justifications," even if those justifications are not their true reasons.

Page 193, but extinction can never be truly undone: There may be an exception to the assertion that extinction can never be undone. For example, if a species' embryos were cryopreserved and it was determined that the embryos could later be viably used, this would create a situation in which no living individuals of the species existed, but they could be readily revived. However, there is a philosophical question worth considering here: if a species is no longer represented by living individuals, but is stored in a way that can be reanimated—such as cryopreserved corals or bucardo cells, or even seeds in a seedbank—is it really extinct? Or does extinction only occur once these stored materials can no longer be revived?

CHAPTER 8

Page 196, escort Napoleon to Saint Helena and set up whatever means were necessary to ensure that he did not escape again: Hart-Davis (2016), p. 23, notes that Cockburn "was in sole charge of the arrangements for Napoleon's security" and that he "did not hesitate to carry out any measure that seemed necessary."

Page 196, "to prevent America or any other nation from planting themselves there": This quote from Cockburn's letter is included in Hart-Davis (2016), p. 23.

Page 197, "You will then take measures to ensure that every boat or vessel which may approach Ascension be minutely examined": Cockburn's written orders for the occupation of Ascension are quoted in Hart-Davis (2016), p. 24.

Page 197, the British slowly began what has become one of the most laborious, and arguably successful, terraforming projects on Earth: Wilkinson (2004) describes the changes made to Ascension Island's plant community as "terraforming."

Page 198, average daily temperatures rising to 95°F (35°C): Duffey (1958) cited an average daily high temperature in Georgetown, Ascension Island, of 35°C, which is approximately 95°F.

Page 198, "The place was of no use": Attributed to Afonso de Albuquerque in 1503 in Hart-Davis (2016), p. 3.

Page 198, "like a land thatt [sic] God has cursed": Attributed to Peter Mundy in 1656 in Hart-Davis (2016), p. 5.

Page 198, "It was a ruinous heap of rocks": Attributed to George Forster, who sailed with Captain James Cook in 1775, in Hart-Davis (2016), p. 19.

Page 198, "One may as easy walk over broken glas [sic] bottles": Attributed to Captain James Cook in 1775 in Hart-Davis (2016), p. 20.

Page 198, "I never saw a more disagreeable place in all the world": Attributed to Peter Osbeck in Duffey (1958).

Page 198, "In all directions, nothing but the most barren and desolate region": Attributed to Reverend. C. I. Latrobe in 1816 in Hart-Davis (2016).

Page 198, "Any body would have believed that the Devil himself had moved his quarters": Attributed to Jan Svilt in Hart-Davis (2016). Svilt was marooned and left a diary on the island before dying, presumably in 1725.

Page 199, the total annual rainfall at only about 5 inches (13 cm): Duffey (1958) reported an average of 5.2 inches of rainfall per year between 1899 and 1954 in Georgetown, Ascension.

Page 199, When it was first sighted by a Portuguese ship captain in 1501: Hart-Davis (2016), p. 1, mentions that sixteenth-century Portuguese explorer João da Nova named the island Conception.

Page 199, Early scientific explorers consistently found only five types of plants: Duffey (1958), p. 225, noted that, "Nearly all the early scientific visitors recorded the presence of the same five plants."

Page 199, "Neither on the sides, nor on the top, did I meet with one single plant": Attributed to Peter Osbeck in Hart-Davis (2016), p. 18.

Page 199, plants did grow atop the very highest peak: Duffey (1958), p. 226, described the precontact top of Green Mountain as probably consisting of "mainly ferns, with mosses and lichens on the moist cool crest of the Green Mountain ridge."

Page 199, Later explorations identified about twenty-five total plant species on the island: Catling and Stroud (2013) noted that, prior to the arrival of people on Ascension, "the total number of plants was probably around 25 to 30, of which ten were endemic: two grasses, two shrubs, and six ferns."

Page 200, Wells dug in the lowlands were salty: Hart-Davis (2016), describes the unsuccessful search for freshwater wells (p. 26) and mentions the higher elevation locations where scant water was found (pp. 10–12).

Page 200, hardy goats that had been introduced to the island: Hart-Davis (2016) describes the introduction of goats that were later used for food, as well as the practice of eating turtles, seabird eggs, and fish.

Page 200, "select the most promising piece of ground . . . and cultivate it": Hart-Davis (2016), p. 29.

Page 200, In this small plot, they planted carrots, turnips, and potatoes: Hart-Davis (2016), pp. 29 and 99.

Page 200, "By then, the island had taken on another role": Hart-Davis (2016), p. 38.

Page 200, In 1843, Ascension received perhaps its most consequential visitor: Hart-Davis (2016), pp. 106–108, describes Hooker's background and visit to Ascension Island.

Page 201, offer suggestions on how to improve the productivity of the island's vegetation and conserve water: As described in Wilkinson (2004).

Page 201, good candidate species, which could be shipped in: Duffey (1958), p. 227, described Hooker's plan for planting Ascension, as well as some of the locations where plants could be sourced.

Page 201, Workers planted eucalyptus trees: Hart-Davis (2016) describes many of the species brought to Ascension in the mid 1800s.

Page 201, One shipment in 1858 included 228 species of plants: Duffey (1958), p. 227, notes that "In 1858 a consignment of plants from the Botanic Gardens at Capetown [sic] included 228 species."

Page 202, He identified 128 species that showed evidence of having naturalized: Duffey (1958), p. 230, noted that "The list of approximately 128 species is therefore very incomplete as it does not include plants which were confined only to gardens, path edges and other human disturbance and settlement areas."

Page 202, In 2013, botanists estimated that from 200 to 300 plant species were growing: Catling and Stroud (2013).

Page 202, The temperature is 72° F (22° C): Duffey (1958) cited an average daily high temperature on Green Mountain of 22.3°C, which is approximately 72°F.

Page 202, This is the top of Green Mountain today: This description of the pool near the peak of Green Mountain is based largely on a photo attributed to Euan Nisbet, in Catling and Stroud (2013), as well as other photos and videos posted on the Internet.

Page 202, "carpet of ferns and here and there a shrub": Duffey (1958), p. 226, quoted Joseph Hooker, from an 1867 article, "On insular floras," published in the *Gardener's Chronicle*.

Page 203, they typically occupy small pockets in tropical mountains: Bruijnzeel et al. (2011).

Page 203, Some ecologists have proposed that this is how ecological communities form: Wilkinson (2004) discusses whether ecosystems are the result of long co-evolution or rather chance events.

Page 204, reproductive rescue to grow its population in uncrowded conditions: In the case of Ascension Island, the plants that survived would have been undergoing reproductive rescue in the sense that they were starting at a low population size and using reproduction to rise toward their carrying capacity.

Page 204, "The consequences of anthropogenic change have been severe" as areas are "infested with undesirable species": Lambdon et al. (2009), pp. 3, 6, 7, 8.

Page 205, some are at risk of extinction without ongoing care and maintenance by island residents and naturalists: Lambdon et al. (2009) provides information about the conservation status of Ascension's native plants.

Page 205, creating novel habitats in which native species can persist among the more recent arrivals: Lambdon et al. (2009), p. 7, argues that truly restoring Ascension's plant communities is impossible, and instead an approach that involves "the creation of 'novel' habitats created from a mix of native species" is most appropriate.

Page 206, author Fred Pearce calls "green xenophobia": Pearce (2015), Introduction, p. vii.

Page 207, some believe that Ascension Island was named by Afonso de Albuquerque: Hart-Davis (2016), pp. 2–3, notes that, according to tradition, Albuquerque named the island on Ascension Day 1503; however, the surviving account of de Albuquerque's sighting of the island seems to indicate that it had been previously so named.

CHAPTER 9

Page 211, fundamental changes to local ecosystems that did not involve human meddling: People were likely involved in the extinction of some Pleistocene species, but the reason for the loss of the steppe ecosystem in Europe was more likely natural climate change as the Pleistocene ended and the warmer Holocene began.

Page 219, lionfish can change the reef by consuming a large number of small fish: Côté et al. (2013) reviewed the scientific research about how lionfish affect Atlantic fish communities and concluded that eradication efforts should continue.

Page 220, with existing technology, we cannot remove all the lionfish from the Caribbean: Perhaps someday people will develop new technologies that permit the removal of species like lionfish, but existing approaches can only reduce their abundance.

Page 221, we stand to lose about 1 percent of our current species within the next century: As mentioned in the notes for the Introduction of this book, De Vos et al. (2015) estimated a background rate of extinction of 0.1 extinction per million species per year and noted that the current extinction rate is roughly 1000 times higher than the background rate. Scientists have already described about two million currently living species, though there may be many millions more yet to be discovered. Considering just the two million described species, the background rate of extinction is about one species extinction globally every five years. Since our current rate of extinction roughly 1000 times higher, we're now losing roughly 200 species per year. This represents a 0.01 percent loss of global species diversity per year. At this rate, it would take a century to lose 1 percent of the known species on the planet.

Page 222, As of March 2022, the Red List documented more than 40,000 species that are threatened with extinction globally: Table 1a on the IUCN Red List website, at www.iucnredlist.org/resources/summary-statistics#Summary%20Tables, summarizes the number of species evaluated in relation to the overall number of described species, and numbers of threatened species by major groups of organisms.

Page 222, they are "facing an extremely high risk of extinction in the wild": As of September 2021, IUCN listed 8404 species as *critically endangered* in Table 2, located at www.iucnredlist.org/resources/summary-statistics#Summary%20Tables. IUCN describes critically endangered species as "facing an extremely high risk of extinction in the wild" on p. 15 of *IUCN Red List Categories and Criteria: Version 3.1*, 2nd edition (2012). Note also that this refers to extinction in the wild, even if those species persist in captivity.

Page 223, a whopping 26 percent of all mammals are threatened with extinction: As of September 2021, IUCN estimated that 26 percent of mammal species are threatened, as shown in on Table 1a on the IUCN Red List website, at www.iucnred list.org/resources/summary-statistics#Summary%20Tables.

Page 223, 42 percent of the threatened mammals on the Red List are subcategorized as vulnerable: As of September 2021, the IUCN Red List included 555 mammal species classified as vulnerable out of 1327 mammal species listed as threatened (a combination of three designations: critically endangered, endangered, and vulnerable), on its website, at www.iucnredlist.org/resources/summary-statistics# Summary%20Tables. Thus, vulnerable mammals make up about 42 percent (555/1327) of the threatened mammals.

Page 224, greater than 10 percent over the next century, the species is classified as vulnerable: The criteria IUCN uses to define a species as vulnerable is discussed on pp. 20–22 of *IUCN Red List Categories and Criteria: Version 3.1*, 2nd edition.

Page 224, only about 3.5 percent of mammals are designated as critically endangered: As of September 2021, the IUCN estimates that 225 species of mammals are currently critically endangered out of a total of 6351 known mammal species (Table 1a and Table 3, at www.iucnredlist.org/resources/summary-statistics), meaning that roughly 3.5 percent of existing mammals are currently designated as critically endangered.

Page 224, their overall risk levels have remained fairly stable in recent decades: The IUCN's Red List Index shows trends in extinction risk. Figure 1, at www.iucnred list.org/assessment/red-list-index, shows that the extinction risk to mammals is low relative to some of the other groups listed, and the risk value changed very little (perhaps increasing slightly) from around 1990 to around 2010. Trends for birds, corals, and amphibians are also shown in this figure.

Page 226, Left to their own devices, many populations will likely have evolved: Leaving housecats to their own devices would require not eradicating them where they have been introduced or subjecting them to ongoing gene flow from people redistributing more cats, which would inhibit local evolution.

Bibliography

INTRODUCTION

Brown, J. H., and A. Kodric-Brown. 1977. Turnover rates in insular biogeography: Effect of Immigration on Extinction. *Ecology*, 58(2): 445–449.

Day, S., and M. Maslin. 2005. Linking large impacts, gas hydrates, and carbon isotope excursions through widespread sediment liquefaction and continental slope failure: The example of the K-T boundary event. Special Paper of the Geological Society of America, 384: 239–258.

De Vos, J. M., L. N. Joppa, J. L. Gittleman, P. R. Stephens, and S. L. Pimm. 2015. Estimating the normal background rate of species extinction. *Conservation Biology*, 29(2): 452–462.

Jablonski, D. 2004. Extinction: Past and present. *Nature*, 427: 589.

Richards, M. A., W. Alvarez, S. Self, L. Karlstrom, et al. 2015. Triggering of the largest Deccan eruptions by the Chicxulub impact. *Bulletin of the Geological Society of America*, 127(11–12): 1507–1520.

Schulte, P., L. Alegret, I. Arenillas, et al. 2010. The Chicxulub asteroid impact and mass extinction at the Cretaceous-Paleogene boundary. *Science*, 327(5970): 1214–1218.

Vellekoop, J., A. Sluijs, J. Smit, et al. 2014. Rapid short-term cooling following the Chicxulub impact at the Cretaceous-Paleogene boundary. *Proceedings of the National Academy of Sciences of the United States of America*, 111: 7537–7541.

CHAPTER 1

Chundawat, R. 2018. *The Rise and Fall of the Emerald Tigers: Ten Years of Research in Panna National Park*. New Delhi: Speaking Tiger Books.

Dinerstein, E., C. Loucks, E. Wikramanayake, et al. 2007. The fate of wild tigers. *BioScience*, 57(6): 508–514.

Dutta, T., S. Sharma, B. H. McRae, P. S. Roy, and R. DeFries. 2016. Connecting the dots: Mapping habitat connectivity for tigers in central India. *Regional Environmental Change*, 16: 53–67.

Jhala, Y. V., Q. Qureshi, and A. K. Nayak. 2020. *Status of Tigers, Copredators and Prey in India*. National Tiger Conservation Authority, Government of India, New Delhi, and Wildlife Institute of India, Dehradun.

Kolipaka, S. S. 2018. Can tigers survive in human-dominated landscapes? Understanding human-tiger coexistence in the bufferzone of Panna Tiger Reserve, India. (Ph.D. thesis, Leiden University).

Kolipaka, S. S., W. L. M. Tamis, M. Van 't Zelfde, G. A. Persoon, and H. H. De Iongh. 2018. New insights into the factors influencing movements and spatial distribution of reintroduced Bengal tigers (*Panthera tigris tigris*) in the human-dominated buffer zone of Panna Tiger Reserve, India. *Mammalia*, 82(3): 207–217.

Luo, S. J, W. E. Johnson, L. Smith, J. L. D. David, and S. J. O'Brien. 2010. "What Is a Tiger? Genetics and Phylogeography." In *Tigers of the World*, 2nd ed. Amsterdam: Elsevier Publishing.

Moyle, B. 2009. The black market in China for tiger products. *Global Crime*, 10(1–2): 124–143.

Rathore, C. S., Y. Dubey, A. Shrivastava, P. Pathak, and V. Patil. 2012. Opportunities of habitat connectivity for tiger (*Panthera tigris*) between Kanha and Pench National Parks in Madhya Pradesh, India. *PLoS ONE* 7(7).

Seidensticker, J., B. Gratwicke, and M. Shrestha. 2010. "How Many Wild Tigers Are There? An Estimate for 2008." In *Tigers of the World*, 2nd ed. Amsterdam: Elsevier Publishing.

Sen, P. K., Q. Qureshi, C. Behera, and S. P. Yadav. 2009. *Report on Disappearance of Tigers from Panna Tiger Reserve*.

Sharma, K., B. Wright, T. Joseph, N. Desai. 2014. Tiger poaching and trafficking in India: Estimating rates of occurrence and detection over four decades. *Biological Conservation*, 179: 33–39.

Singh, R., P. Nigam, Q. Qureshi, K. Sankar, P. R. Krausman, and S. P. Goyal. 2014. Strategy of female tigers to avoid infanticide. *Current Science*, 107(9): 1595–1597.

Sunquist, M. 2010. "What Is a Tiger? Ecology and Behavior." In *Tigers of the World*, 2nd ed. Amsterdam: Elsevier Publishing.

Wright, B. 2010. "Will the tiger survive in India?" In *Tigers of the World*, 2nd ed. Amsterdam: Elsevier Publishing.

CHAPTER 2

Carothers, C., T. L. Sformo, S. Cotton, J. C. George, and P. A. H. Westley. 2019. Pacific salmon in the rapidly changing arctic: Exploring local knowledge and emerging fisheries in Utqiaġvik and Nuiqsut, Alaska. *Arctic*, 72(3): 273–288.

Beacham, T. D., B. McIntosh, C. MacConnachie, et al. 2006. Pacific Rim population structure of sockeye salmon as determined from microsatellite analysis. *Transactions of the American Fisheries Society*, 135(1): 174–187.

Kaeriyama, M., M. Nakamura, M. Yamaguche, et al. 2000. Feeding ecology of sockeye and pink salmon in the Gulf of Alaska. *North Pacific Anadromous Fishery Committee Bulletin*, 2: 55–63.

Quinn, T. P., L. Wetzel, S. Bishop, K. Overberg, and D. E. Rogers. 2001. Influence of breeding habitat on bear predation and age at maturity and sexual dimorphism of sockeye salmon populations. *Canadian Journal of Zoology*, 79: 1782–1793.

Schindler, D. E., R. Hilborn, B. Chasco, et al. 2010. Population diversity and the portfolio effect in an exploited species. *Nature*, 465: 609–612.

CHAPTER 3

Archer, M., H. Bates, S. J. Hand, et al. 2019. The Burramys project: A conservationist's reach should exceed history's grasp, or what is the fossil record for? *Philosophical Transactions of the Royal Society B: Biological Sciences*, 374.

Broom, R. 1896. On a small fossil marsupial with large grooved premolars. *Proceedings of the Linnean Society of New South Wales*, 10: 563–567.

Broome, L. S. 2001. Density, home range, seasonal movements and habitat use of the mountain pygmy-possum *Burramys parvus* (Marsupialia: Burramyidae) at Mount Blue Cow, Kosciuszko National Park. *Austral Ecology*, 26(3): 275–292.

Broome, L., M. Archer, H. Bates, H. Shi, et al. 2012. "A brief review of the life history of, and threats to, *Burramys parvus* with a prehistory-based proposal for ensuring that it has a future." In *Wildlife and Climate Change: Toward robust conservation strategies for Australian fauna*. New South Wales, Australia: Royal Zoological Society of New South Wales.

Broome, L., D. Heinze, and M. Schroder. 2018. "Recovering the mountain pygmy-possum at Mt. Blue Cow and Mt. Buller." In *Recovering Australian Threatened Species: A Book of Hope*. Clayton, Victoria, Australia: CSIRO Publishing.

Department of Environment, Land, Water and Planning. 2016. *National Recovery Plan for the Mountain Pygmy-possum* Burramys parvus. Australian Government, Canberra.

Glen, A. S., A. E. Byrom, R. P. Pech, et al. 2012. Ecology of brushtail possums in a New Zealand dryland ecosystem. *New Zealand Journal of Ecology*, 36(1): 29–37.

Green, K., P. Caley, M. Baker, D. Dreyer, J. Wallace, and E. Warrant. 2020. Australian Bogong moths *Agrotis infusa* (Lepidoptera: Noctuidae), 1951–2020: decline and crash. *Austral Entomology*, 60(1): 66–81.

Gynther, I., N. Waller, and L. K. Leung. 2016. *Confirmation of the extinction of the Bramble Cay melomys* Melomys rubicola *on Bramble Cay, Torres Strait: Results and conclusions from a comprehensive survey in August–September 2014*. Report to the Department of Environment and Heritage Protection, Queensland Government, Brisbane.

Meredith, R. W., M. Westerman, J. A. Case, and M. S. Springer. 2008. A phylogeny and timescale for marsupial evolution based on sequences for five nuclear genes. *Journal of Mammalian Evolution*, 15: 1–36.

Ride, D. L. 1956. The affinities of *Burramys parvus* Broom—a fossil Phalangeroid marsupial. *Journal of Zoology*, 127(3): 413–429.

Stephenson, B., B. David, J. Fresløv, et al. 2020. 2000 Year-old Bogong moth (*Agrotis infusa*) Aboriginal food remains, Australia. *Scientific Reports*, 10: 1–10.

Thomas, C. D. 2017. *Inheritors of the Earth: How Nature Is Thriving in an Age of Extinction*. New York: PublicAffairs.

Woinarski, J. C. 2016. A very preventable mammal extinction. *Nature*, 535: 493.

CHAPTER 4

Anagnostakis, S. L. 1987. Chestnut blight: The classical problem of an introduced pathogen. *Mycologia*, 79(1): 23–37.

Barakat, A., M. Staton, C. H. Cheng, J. Park, et al. 2012. Chestnut resistance to the blight disease: Insights from transcriptome analysis. *BMC Plant Biology*, 12: 38.

Brewer, L. G. 1995. Ecology of survival and recovery from blight in American chestnut trees (*Castanea dentata* (Marsh.) Borkh.) in Michigan. *Bulletin of the Torrey Botanical Club*, 122(1): 40–57.

Dalgleish, H. J., C. D. Nelson, J. A. Scrivani, and D. F. Jacobs. 2016. Consequences of shifts in abundance and distribution of American chestnut for restoration of a foundation forest tree. *Forests*, 7(1): 1–9.

Deng, T., Z. Nie, B. T. Drew, S. Volis, and C. Kim. 2015. Does the Arcto-Tertiary biogeographic hypothesis explain the disjunct distribution of Northern Hemisphere herbaceous plants ? The case of Meehania (Lamiaceae). *PLoS ONE*, 10(2): 1–18.

Double, M. L., A. M. Jarosz, D. W. Fulbright, A. Davelos Baines, and W. L. MacDonald. 2018. Evaluation of two decades of *Cryphonectria parasitica* hypovirus Introduction in an American Chestnut Stand in Wisconsin. *Phytopathology*, 108(6): 702–710.

Freinkel, S. 2007. *American Chestnut: The Life, Death, and Rebirth of a Perfect Tree.* Berkeley and Los Angeles: University of California Press.

Gravatt, G. F. 1925. "The Chestnut Blight in North Carolina." In *Chestnut and the Chestnut Blight in North Carolina.* North Carolina Geologic and Economic Survey, Economic Paper, 56: 13–17.

Kyndt, T., D. Quispe, H. Zhai, et al. 2015. The genome of cultivated sweet potato contains *Agrobacterium* T-DNAs with expressed genes: An example of a naturally transgenic food crop. *Proceedings of the National Academy of Sciences of the United States of America*, 112(18): 5844–5849.

Lang, P., F. Dane, and T. L. Kubisiak. 2006. Phylogeny of Castanea (Fagaceae) based on chloroplast trnT-L-F sequence data. *Tree Genetics & Genomes*, 2: 132–139.

Lang, P., F. Dane, T. L. Kubisiak, and H. Huang. 2007. Molecular evidence for an Asian origin and a unique westward migration of species in the genus *Castanea* via Europe to North America. *Molecular Phylogenetics and Evolution*, 43(1): 49–59.

Li, F. W., J. C. Villarreal, S. Kelly, et al. 2014. Horizontal transfer of an adaptive chimeric photoreceptor from bryophytes to ferns. *Proceedings of the National Academy of Sciences of the United States of America*, 111: 6672–6677.

Merkel, H. 1906. A deadly fungus on the American chestnut. *Annual Report of the New York Zoological Society*, 10: 97–103.

Moran, N. A., and T. Jarvik. 2010. Lateral transfer of genes from fungi underlies carotenoid production in aphids. *Science*, 328(5978): 624–627.

Murrill, W. 1908. The chestnut canker. *Torreya: A Monthly Journal of Botanical Notes and News*, 8: 111–112.

Newhouse, A. E., V. C. Coffey, et al. 2020. State University of New York College of Environmental Science and Forestry; Petition for determination of nonregulated status for blight-resistant Darling 58 American chestnut. (Petition 19-309-01p_a1). *Federal Register*, 85(161): 51008–51009.

Novak, B., O. Ryder, B. Wiese, et al. 2016. A proposal for genomically adapting black-footed ferrets for disease immunity. Proposal submitted to the US Fish & Wildlife Service.

Opler, P. A. 1979. Insects of American chestnut: Possible importance and conservation concern. *American Chestnut Proceedings*, 83–85.

Ow, D. W., J. R. D. E. Wet, D. R. Helinski, et al. 1986. Transient and stable expression of the firefly luciferase gene in plant cells and transgenic plants. *Science*, 234(4778): 856–859.

Powell, W. A., A. E. Newhouse, and V. Coffey. 2019. Developing blight-tolerant American chestnut trees. *Cold Spring Harbor Perspectives in Biology.*

Prospero, S., M. Conedera, U. Heiniger, and D. Rigling. 2006. Saprophytic activity and sporulation of *Cryphonectria parasitica* on dead chestnut wood in forests with naturally established hypovirulence. *Phytopathology,* 96(12): 1337–1344.

Steiner, K. C., J. W. Westbrook, F. V. Hebard, et al. 2017. Rescue of American chestnut with extraspecific genes following its destruction by a naturalized pathogen. *New Forests,* 48: 317–336.

Westbrook, J. W., J. A. Holliday, A. E. Newhouse, and W. A. Powell. 2020. A plan to diversify a transgenic blight-tolerant American chestnut population using citizen science. *Plants People Planet,* 2(1): 84–95.

CHAPTER 5

Bayraktarov, E., P. J. Stewart-Sinclair, S. Brisbane, et al. 2019. Motivations, success, and cost of coral reef restoration. *Restoration Ecology,* 27(5): 981–991.

Burke, L., K. Reytar, M. Spalding, and A. Perry. 2011. *Reefs at Risk Revisited.* World Resources Institute.

Charteris, M. 2018. *Caribbean Reef Life: A Field Guide for Divers,* digital 3rd ed.

DeFilippo, L. B., L. C. McManus, D. E. Schindler, et al. 2022. Assessing the potential for demographic restoration and assisted evolution to build climate resilience on coral reefs. *Ecological Applications.*

Gardner, T. A., I. M. Côté, J. A. Gill, A. Grant, and A. R. Watkinson. 2003. Long-term region-wide declines in Caribbean corals. *Science,* 301(5635): 958–960.

Hoegh-Guldberg, O., L. Pendleton, and A. Kaup. 2019. People and the changing nature of coral reefs. *Regional Studies in Marine Science,* 30: 100699.

Hughes, T. P., K. D. Anderson, S. R. Connolly, et al. 2018. Spatial and temporal patterns of mass bleaching of corals in the Anthropocene. *Science,* 359(6371): 80–83.

Knowlton, N., R. E. Brainard, R. Fisher, et al. 2010. "Coral Reef Biodiversity." In *Life in the World's Oceans: Diversity, Distribution, and Abundance.* Oxford, UK: Wiley-Blackwell Publishing.

Muller, E. M., C. Sartor, N. I. Alcaraz, and R. van Woesik. 2020. Spatial epidemiology of the stony-coral-tissue-loss disease in Florida. *Frontiers in Marine Science,* 7: 163.

Palumbi, S. R., D. J. Barshis, N. Traylor-Knowles, and R. A. Bay. 2014. Mechanisms of reef coral resistance to future climate change. *Science,* 344(6186): 895–898.

Riul, P., C. H. Targino, L. A. C. Júnior, et al. 2013. Invasive potential of the coral *Tubastraea coccinea* in the southwest Atlantic. *Marine Ecology Progress Series,* 480: 73–81.

Sully, S., D. E. Burkepile, M. K. Donovan, G. Hodgson, and R. van Woesik. 2019. A global analysis of coral bleaching over the past two decades. *Nature Communications*, 10: 1–5.

Taylor, J. E., M. Filipski, A. Bond Manuso, et al. 2011. *Evaluación de Impactos Ambientales y Socioeconómicos Programa de Manejo Ambiental de Islas de la Bahía-Fase II*. Washington, D.C. and Tegucigalpa: Inter-American Development Bank.

van Oppen, M. J. H., J. K. Oliver, H. M. Putnam, and R. D. Gates. 2015. Building coral reef resilience through assisted evolution. *Proceedings of the National Academy of Sciences of the United States of America*, 112: 2307–2313.

Walsworth, T. E., D. E. Schindler, M. A. Colton, et al. 2019. Management for network diversity speeds evolutionary adaptation to climate change. *Nature Climate Change* 9: 632–636.

Zanna, L., S. Khatiwala, J. M. Gregory, J. Ison, and P. Heimbach. 2019. Global reconstruction of historical ocean heat storage and transport. *Proceedings of the National Academy of Sciences of the United States of America*, 116(4): 1126–1131.

CHAPTER 6

Barel, C., R. Dorit, P. Greenwood, et al. 1985. Destruction of fisheries in Africa's lakes. *Nature*, 315: 19–20.

Chorowicz, J. 2005. The East African rift system. *Journal of African Earth Sciences*, 43(1–3): 379–410.

Fryer, G. 1960. Concerning the proposed introduction of Nile perch into Lake Victoria. *The East African Agricultural Journal*, 25(4): 267–270.

Goldschmidt, T. 1996. *Darwin's Dreampond: Drama in Lake Victoria*, 2nd ed. Translated by Sherry Marx-Macdonald. Cambridge, MA: The MIT Press.

Johnson, T. C., K. Kelts, and E. Odada. 2000. The holocene history of Lake Victoria. *Ambio: A Journal of the Human Environment*, 29(1): 2–11.

Katunzi, E. F. B., J. Zoutendijk, T. Goldschmidt, J. H. Wanink, and F. Witte. 2003. Lost zooplanktivorous cichlid from Lake Victoria reappears with a new trade. *Ecology of Freshwater Fish*, 12(4): 237–240.

Kaufman, L. 1992. Catastrophic change in species-rich freshwater ecosystems. *BioScience*, 42(11): 846–858.

Kishe-Machumu, M. A., J. C. van Rijssel, J. H. Wanink, and F. Witte. 2015. Differential recovery and spatial distribution pattern of haplochromine cichlids in the Mwanza Gulf of Lake Victoria. *Journal of Great Lakes Research*, 41(2): 454–462.

Marques, D. A., J. I. Meier, and O. Seehausen. 2019. A combinatorial view on speciation and adaptive radiation. *Trends in Ecology & Evolution*, 34(6): 531–544.

Marshall, B. E. 2018. Guilty as charged: Nile perch was the cause of the haplochromine decline in Lake Victoria. *Canadian Journal of Fisheries and Aquatic Sciences*, 75(9): 1542–1559.

Matsuishi, T., L. Muhoozi, O. Mkumbo, et al. 2006. Are the exploitation pressures on the Nile perch fisheries resources of Lake Victoria a cause for concern? *Fisheries Management and Ecology*, 13(1): 53–71.

McGee, M. D., S. R. Borstein, J. I. Meier, et al. 2020. The ecological and genomic basis of explosive adaptive radiation. *Nature*, 586: 75–79.

McGee, M. D., S. R. Borstein, R. Y. Neches, et al. 2015. A pharyngeal jaw evolutionary innovation facilitated extinction in Lake Victoria cichlids. *Science*, 350(6264): 1077–1079.

Meier, J. I., D. A. Marques, S. Mwaiko, et al. 2017a. Ancient hybridization fuels rapid cichlid fish adaptive radiations. *Nature Communications*, 8: 1–11.

Meier, J. I., V. C. Sousa, D. A. Marques, et al. 2017b. Demographic modelling with whole-genome data reveals parallel origin of similar *Pundamilia* cichlid species after hybridization. *Molecular Ecology*, 26(1): 123–141.

Moser, F. N., J. C. van Rijssel, S. Mwaiko, et al. 2018. The onset of ecological diversification 50 years after colonization of a crater lake by haplochromine cichlid fishes. *Proceedings of the Royal Society B: Biological Sciences*, 285(1884): 20180171.

Mzighani, S. I., M. Nikaido, M. Takeda, et al. 2010. Genetic variation and demographic history of the *Haplochromis laparogramma* group of Lake Victoria: An analysis based on SINEs and mitochondrial DNA. *Gene*, 450(1–2): 39–47.

Njiru, J., M. van der Knaap, R. Kundu, and C. Nyamweya. 2018. Lake Victoria fisheries: Outlook and management. *Lakes & Reservoirs Research & Management*, 23(2): 152–162.

Odongkara, K., R. Abila, and J. Luomba. 2010. The contribution of Lake Victoria fisheries to national economies. *African Journal of Tropical Hydrobiology and Fisheries*, 12(1).

Phillips, B. L., and R. Shine. 2006. An invasive species induces rapid adaptive change in a native predator: Cane toads and black snakes in Australia. *Proceedings of the Royal Society B: Biological Sciences*, 273(1593): 1545–1550.

Pringle, R. M. 2005. The origins of the Nile perch in Lake Victoria. *BioScience*, 55(9): 780–787.

Quinn, T. P., M. T. Kinnison, and M. J. Unwin. 2001. Evolution of chinook salmon (*Oncorhynchus tshawytscha*) populations in New Zealand: Pattern, rate, and process. *Genetica*, 112–113(1): 493–513.

Seehausen, O., J. J. M. Van Alphen, and F. Witte. 1997. Cichlid fish diversity threatened by eutrophication that curbs sexual selection. *Science*, 277(5333): 1808–1811.

Thomas, C. D. 2015. Rapid acceleration of plant speciation during the Anthropocene. *Trends in Ecology and Evolution*, 30(8): 448–455.

van Rijssel, J. C., and F. Witte. 2013. Adaptive responses in resurgent Lake Victoria cichlids over the past 30 years. *Evolutionary Ecology*, 27(2): 253–267.

Witte, F., C. D. N. Barel, E. L. M. Witte-Maas, and M. J. P. van Oijen. 1976. An introduction to the taxonomy and morphology of the haplochromine Cichlidae from Lake Victoria. *Netherlands Journal of Zoology*, 27(4): 333–380.

Witte, F., T. Goldschmidt, J. Wanink, et al. 1992. The destruction of an endemic species flock: quantitative data on the decline of the haplochromine cichlids of Lake Victoria. *Environmental Biology of Fishes*, 34(1): 1–28.

Witte, F., B. S. Msuku, J. H. Wanink, et al. 2000. Recovery of cichlid species in Lake Victoria: An examination of factors leading to differential extinction. *Reviews in Fish Biology and Fisheries*, 10: 233–241.

Witte, F., and E. L. M. Witte-Maas. 1987. "Implications for Taxonomy and Functional Morphology of Intraspecific Variation in Haplochromine Cichlids of Lake Victoria with descriptions of five zooplanktivorous species." In *From Form to Fishery: an ecological and taxonomical contribution to morphology and fishery of Lake Victoria cichlids*. (Ph.D. thesis, Leiden University).

CHAPTER 7

Barlow, A., J. L. A. Paijmans, F. Alberti, et al. 2021. Middle Pleistocene genome calibrates a revised evolutionary history of extinct cave bears. *Current Biology*, 31(8): 1771–1779.

Beer, C., N. Zimov, J. Olofsson, P. Porada, and S. Zimov. 2020. Protection of permafrost soils from thawing by increasing herbivore density. *Scientific Reports* 10(1): 4170.

Borges, A. A., and A. F. Pereira. 2019. Potential role of intraspecific and interspecific cloning in the conservation of wild mammals. *Zygote*, 27(3): 111–117.

Calef, G. W. 1988. Maximum rate of increase in the African elephant. *African Journal of Ecology*, 26(4): 323–327.

Chauvet, J., E. B. Deschamps, and C. Hillaire. 1996. *Dawn of Art: The Chauvet Cave: The Oldest Known Paintings in the World*. New York: Harry N. Abrams.

Christiansen, P. 1999. What size were *Arctodus simus* and *Ursus spelaeus* (Carnivora: Ursidae)? *Annales Zoologici Fennici*, 36(2): 93–102.

Döppes, D., G. Rabeder, C. Frischauf, et al. 2019. Extinction pattern of Alpine cave bears - new data and climatological interpretation. *Historical Biology*, 31(4): 422–428.

Folch, J., M. J. Cocero, P. Chesné, et al. 2009. First birth of an animal from an extinct subspecies (*Capra pyrenaica pyrenaica*) by cloning. *Theriogenology*, 71(6): 1026–1034.

Grayson, D. K., and F. Delpech. 2003. Ungulates and the Middle-to-Upper Paleolithic transition at Grotte XVI (Dordogne, France). *Journal of Archaeological Science*, 30: 1633–1648.

Kunda, Z. 1990. The case for motivated reasoning. *Psychological Bulletin*, 108(3): 480–498.

MacPhee, R. D. E. 2018. *End of the Megafauna: The Fate of the World's Hugest, Fiercest, and Strangest Animals*. New York: W. W. Norton & Company.

Mondanaro, A., M. Di Febbraro, M. Melchionna, et al. 2019. Additive effects of climate change and human hunting explain population decline and extinction in cave bears. *Boreas*, 48(3): 605–615.

Pacher, M., and A. J. Stuart. 2009. Extinction chronology and palaeobiology of the cave bear (*Ursus spelaeus*). *Boreas*, 38(2): 189–206.

Shapiro, B. 2015. *How to Clone a Mammoth: The Science of De-Extinction*. Princeton, NJ: Princeton University Press.

Stiller, M., G. Baryshnikov, H. Bocherens, et al. 2010. Withering away–25,000 years of genetic decline preceded cave bear extinction. *Molecular Biology and Evolution*, 27(5): 975–978.

Terlato, G., H. Bocherens, M. Romandini, et al. 2019. Chronological and isotopic data support a revision for the timing of cave bear extinction in Mediterranean Europe. *Historical Biology*, 31: 474–484.

Vartanyan, S. L., K. A. Arslanov, T. V. Tertychnaya, and S. B. Chernov. 1995. Radiocarbon dating evidence for mammoths on Wrangel Island, Arctic Ocean, until 2000 BC. *Radiocarbon*, 37(1): 1–6.

Wilmut, I., A. E. Schnieke, J. Mcwhir, A. J. Kind, and K. H. S. Campbell. 1997. Viable offspring derived from fetal and adult mammalian cells. *Nature*, 385: 810–813.

Zimov, S. A. 2005. Pleistocene park: Return of the mammoth's ecosystem. *Science*, 308(5723): 796–798.

Zimov, S. A., N. S. Zimov, and F. S. C. III. 2012. "The Past and Future of the Mammoth Steppe Ecosystem." In *Paleontology in Ecology and Conservation*, Julien Louys, ed. Berlin, Heidelberg: Springer.

CHAPTER 8

Bruijnzeel, L. A., F. N. Scatena, and L. S. Hamilton. 2011. *Tropical Montane Cloud Forests*. Cambridge, UK: Cambridge University Press.

Catling, D. C., and S. Stroud. 2013. "The Greening of Green Mountain, Ascension Island." In *Post-Sustainable: Blueprints for a Green Planet*, M. Joachim and M. Silver, eds. New York: Metropolis Books.

Duffey, E. 1958. The terrestrial ecology of Ascension Island. *Journal of Applied Ecology*, 1(2): 219–251.

Hart-Davis, D. 2016. *Ascension: the Story of a South Atlantic Island*. Shopshire, UK: Merlin Unwin Books Ltd.

Lambdon, P., S. Sroud, C. Clubbe, et al. 2009. A plan for the conservation of endemic and native flora on Ascension Island. Paper for the Ascension Island Threatened Plants Restoration Project.

Pearce, F. 2015. *The New Wild: Why Invasive Species Will Be Nature's Salvation*. Boston: Beacon Press.

Wilkinson, D. M. 2004. The parable of Green Mountain: Ascension Island, ecosystem construction and ecological fitting. *Journal of Biogeography*, 31(1): 1–4.

CHAPTER 9

Côté, I. M., S. J. Green, and M. A. Hixon. 2013. Predatory fish invaders: Insights from Indo-Pacific lionfish in the western Atlantic and Caribbean. *Biological Conservation*, 164: 50–61.

IUCN. 2012. *IUCN Red List Categories and Criteria: Version 3.1*, 2nd ed. Gland, Switzerland and Cambridge, UK: IUCN Species Survival Commission.

Acknowledgments

Special thanks go to a few individuals who made particularly large contributions to *The Rescue Effect*. My wife, Avani Mehta Sood, encouraged me to take the risk of writing this book, while supporting me as I searched for my voice and developed my chapters; she was my tireless sounding board for ideas—good and bad—and more than any anyone else, she helped me shape my ideas into *The Rescue Effect*. Mat Squillante gave me the best single piece of writing advice I received: write the book as a collection of stories, not a textbook. Piali Mukhopadhyay helped me greatly improve my writing by wading through many early drafts and giving me insightful feedback. My agent, Jill Marsal, and my editor, Will McKay, took a chance on me and supported me throughout the writing and publishing processes. And my children, Quinn, Ishaan, and Samara, were inordinately patient and supportive while my attention was focused on research and writing.

In researching and writing *The Rescue Effect*, I relied on the knowledge and experience of a wide array of experts. I benefited immeasurably from scientists, journalists, historians, and others who took the time to share their knowledge and experiences with the world in the articles, books, databases, management plans, and other media that are cited in the notes and bibliography. I am also deeply indebted to the scores of experts who shared their time, expertise, and personal stories with me through interviews. These include Michael Archer, Vijay Bahn Singh, Axel Barlow, Bridget Baumgartner, Tony Birdsey, Irma Brady, Linda Broome, Mark Carr, Hans Dalal, Herbert Darling, Ramon de Leon, Lukas DiFilippo, Ian Drysdale, Sara Fern Fitzsimmons, Tripp Funderburk, Mary Hagedorn, Ray Hilborn, Les Kaufman, Nancy Knowlton, Shekhar Kolipaka, Matt McGee, Joana Isabel Meier, Bhavna Menon, Tino Monterroso, R. Sreenavasa Murthy, Jenny Myton, Angela Nankabirwa, Richard Ogutu-Ohwayo, Horace Owiti Onyango, Marissa Parrott, Anne Petermann, Ryan Phelan, Malin Pinsky, Joe Pollock, William Powell, Thomas Quinn, Robert Richmond, Andrea Rivera, Bittu Sahgal, Daniel Schindler, Ole Seehausen, Beth Shapiro, Ken Shortman, Neel Soni, Lisa Thomson, Madeline van Oppen, Jacco van Rijssel, Robert Warneke, Virginia Weiss, Jared Westbrook, and Sergey Zimov. Although I was not able to include direct content from all the interviews that I conducted, each and every one of them informed my ideas and conclusions. Thank you all for all your contributions and generosity.

Many people generously reviewed and gave me feedback on earlier drafts and chapters, including Ian Drysdale, Kathryn Fiorella and her lab group at Cornell University, Sara Fern Fitzsimmons, Sarah Freiermuth, Olivia Greenburg, Elizabeth Kast, Jill Marsal, Aruna Mehta, Akash Mehta, Bhavna Menon, Piali Mukhopadhyay, Jenny Myton, Angela Nankabirwa, Horace Owiti Onyango, Marissa Parrott, Thomas Quinn, Daniel Schindler, Mat Squillante, Lisa Thomson, Robert Warneke, and Jared Westbrook. In addition, Cliff Kraft, my editor Will McKay, Timber Press editors Mike Dempsey and Lisa Theobald, and Avani Mehta Sood reviewed the manuscript in its entirety. Thank you all for your invaluable feedback—the final product is much improved because of it. I also thank those individuals who gave me general advice about how to approach writing and publishing this book, including Nancy Baron, Julie Bennett, Charles Conn, Dave Ebert, Amelia Glynn, Allison Hanes, Jennifer Jacquet, Dale Jamieson, Nancy Kelton, Jill Marsal, and Monika Verma.

To everyone who has contributed to the completion of this book: thank you, I couldn't have done this without you.

Index

Japanese chestnut trees, 90, 92
Jardines de la Reina (Gardens of the
    Queen), 212–213
Jurassic Park, 188

K

Kanha National Park (India), 32, 33
Kaufman, Les, 158, 167, 168
Ken River (India), 30
Kenya, 252, 253
Kenya Marine and Fisheries Research
    Institute, 159, 166
Kodiak Island (Alaska), 236
Kosciuszko, Mount (Australia), 63,
    67, 239
Kosciuszko National Park (Australia),
    72
Kuskokwim River (Alaska), 54–55

L

Lake Aleknagik, 44, 45–46, 51
Lake Chala, 171
Lake Malawi, 146
Lake Tanganyika, 146
Lake Victoria. *See* Victoria, Lake
larval corals, behavior of, 120
Last Wilderness Foundation, 28
leopards, 18
Lewis, Carl, 135
life on Earth, history of, 12
ligers, 98–99
lionfish, 219–220, 227, 264
Little Higginbotham, Mount
    (Australia), 72
livestock, 22–23
long-tailed pygmy-possum, 61–62
longwing butterflies, 152

M

Madhya Pradesh (Indian state), 35, 37,
    233
magnetic field, of Earth, 45
Malawi, Lake, 146
mammals
    in Australia, 61–62, 68
    risk of extinction, 222–224, 225, 265
mapping software, wildlife corridors
    and, 36
marine reserves, 125–126, 127
masked owls, 58, 236–237
mass extinction events, 253. *See also*
    global mass extinction events;
    species extinctions
maternal mouthbrooders, 149–150,
    151–152, 249
Maynard, Charles, 101, 103
McGee, Matt, 157, 158, 165, 169
medicines, and coral reefs, 121
megafauna extinctions, 176–179
Meier, Joana Isabel, 147, 148, 153
Menon, Bhavna, 28, 31, 34
Merkel, Hermann, 84
Mesoamerican Reef Data Explorer, 247
Mesoamerican Reef region, 131–132
Mesozoic Era, 13, 88–89
meteors, 13
mimicry in nature, 245
modeling studies, on corals and coral
    reefs, 134–135
Monash University (Australia), 157, 165
monk seals, 211, 212
Monterroso, Tino, 125
moon phases, and coral reproduction,
    119–120
moths, 242. *See also* Bogong moths
motivated reasoning phenomenon, 192

mountain pygmy-possums. *See*
   pygmy-possums
mountains, and geographic rescue of
   pygmy-possums, 75–80
Mount Buller (Australia), 63, 67, 75, 239
Mount Hotham (Australia), 63, 72
Mount Kosciuszko (Australia), 63,
   67, 239
Mount Little Higginbotham
   (Australia), 72
Murrill, William A., 85
Murthy, R. Sreenivasa, 21–22, 30, 32,
   34, 39
museum, nature as, 209–213, 214, 226
mushrooms, 91
Mwanza Gulf (Lake Victoria)
   after Nile perch introduction, 252
   cichlid species decline in, 157
   haplochromine cichlids in, 160–162,
      255
   haplochromine hybrids in, 163–164,
      165, 170, 253, 254
mycobacterial infections, 167
Myton, Jenny, 128, 132, 142

N

Namenalala, Fiji (island), 9
Napoleon Bonaparte. *See* Bonaparte,
   Napoleon
Native Americans, and chestnuts, 87,
   242
native species, 204–205, 209, 241
natural rescue, for corals and coral reefs,
   127–132
natural selection, and salmon spawning
   sites, 49–50
nature-as-museum model, 209–213,
   214, 226

nature parks, 34, 209–210
nature tourism, 39
New Guinea, 59
New York Times, on Bogong moths,
   65–66
New York Zoological Park, 84
New Zealand, 76–77, 80, 255–256
Nile perch. *See* giant Nile perch
Nile River, 147, 154
Nisbet, Euan, 263
Non-GMO Project, 108, 246
Noradehi Wildlife Sanctuary, 37, 233
North American Chinook salmon, 169
Nottingham Trent University, 184
novelty and novel ecosystems, 204,
   206–207, 219–221, 226–227, 263
nuclear transfer process, 181

O

oak trees, 93
octopuses, 116
Ogutu-Ohwayo, Richard, 158, 159
opossums vs. possums, 59, 237
orange cup corals, 139, 248
orangeheads (haplochromine cichlid
   species), 160–162, 163, 253–254
outbreeding depression, and salmon
   strays, 53
Owiti Onyango, Horace, 159, 166
owls, masked, 58, 236–237
oxalate oxidase gene, 104–105, 106,
   107, 108, 245

P

Pacific Ocean, salmon in, 44
paedophages, 149–150, 249
pandas, 165
Panna City, India, 31